Y0-BSA-501

Sky Riders

Sky Riders

An Illustrated History
of Aviation in Alberta
1906-1945

Patricia A. Myers

**FIFTH
HOUSE**
PUBLISHERS

Friends of Reynolds – Alberta
Museum Society

Front cover photograph, "Gladys Walker and 'Wop' May in their flying clothes,"
    courtesy City of Edmonton Archives EA-146-1
Back cover photograph, "Captain Fred McCall with a Curtiss Jenny, 1919," courtesy
    Glenbow–Alberta Institute NA-1258-22
Cover design by John Luckhurst/GDL
Layout and design by Donald Ward/Ward Fitzgerald editorial design

The publisher gratefully acknowledges support received from The Canada Council,
Heritage Canada, and the Saskatchewan Arts Board.

Printed and bound in Canada
95 96 97 98 99 / 5 4 3 2 1

Canadian Cataloguing in Publication Data

Myers, Patricia, 1956–

    Sky riders

    Includes bibliographical references and index.
    ISBN 1-895618-57-6

1. Aeronautics - Alberta - History. I. Title.

TL523.M94 1995   629.13'0097123'09   C95-920045-2

FIFTH HOUSE LTD.
620 Duchess Street
Saskatoon, SK   S7K 0R1

Contents

Foreword

On 17 September 1930, the National Air Tour for the Edsel B. Ford Trophy touched down in Edmonton. I was nine years old. Wilfrid Reid "Wop" May, a family friend, took me by the hand and together we walked around each magnificent plane. When we got to the Texaco Mystery Ship NC1313, Wop lifted me up to the pilot, Captain Frank Hawks, and said, "Some day, this boy's going to be a pilot, too." I did become a pilot, serving with the RCAF during World War II, and then as a pilot, chief pilot, and aviation manager with a number of companies for more than forty years.

My love of flying was nurtured in Alberta, on Wop May's shoulders and at Blatchford Field in Edmonton. I had my first airplane ride in 1933 in a Cirrus Moth that belonged to the Edmonton and Northern Alberta Aero Club. That flight started a passionate love affair with aviation that continues to this day.

During the late 1930s, I was an "airport kid," spending every minute I could in the hangars. I was eager to help with the airplanes, and thrilled to be around the pilots I admired. More recently, I have become involved in preserving many areas of Alberta's past. One area is, of course, this province's exciting history of aviation. I am pleased to have been able to make some small contribution to this illustrated history of flying in Alberta. It is a story that needed telling.

I hope this book brings back memories for many of you, and in others, sparks an interest in what I think is one of life's greatest pursuits—flying.

J. H. "Jack" Reilly
Member, Canada's Aviation Hall of Fame

Acknowledgements

There are few people who are not fascinated by flying, I've discovered. While writing *Sky Riders*, I found that virtually everyone I spoke to had a treasured anecdote about a pilot, an event, or a plane. Such incidents made writing this history of early flying in Alberta a pleasure.

That pleasure was increased by the many people who helped at every stage of the project. This book could not have been completed without the constant support and encouragement of the Friends of Reynolds–Alberta Museum Publication Committee, and I owe a special debt to each member. Under the enthusiastic chairmanship of Jack Manson, the committee assisted with research, read and commented on the manuscript, and always managed to have a word of encouragement. Jack brought not only his tremendous energy to the project, but also his memories of flying days with the cadets and the RCAF, and a detailed knowledge of the printing business. Stan Reynolds made his vast personal collection of research materials available to me. He made detailed comments on the manuscript, and was always ready to answer questions. His recollections were invaluable. Byron Reynolds loaned books from his personal collection to me, and carefully read and commented on each chapter. Jack Reilly shared his memories, suggested improvements to the text, and encouraged me throughout its writing. Lori Marshall, Executive Director of the Friends, helped keep the project moving forward, as did David Dusome.

At Historic Sites and Archives Service, Alberta Community Development, Michael Payne, Frits Pannekoek, and Sandra Thomson supported the project and the author with enthusiasm and determination. Lynn Pong carefully processed the annotated references.

Barbara Hagensen, Steve Boddington, Ken Tingley, and Sheila Tingley helped with aspects of the research. Keith Spencer shared his personal collection of first flight covers with me, as well as his enthusiasm for early mail flights, cachets, and stamps in the province. Bob Hesketh showed me essays from his forthcoming collection on defence projects in northwestern Canada during World War II. Jim Light and Don Wetherell gave me access to family papers. Don also read and commented on the manuscript.

Marlena Wyman of the Provincial Archives of Alberta, and June Honey of the City of Edmonton Archives responded with efficiency and good humour every time I asked for "just one more photo." Dennis Hyduk of the Provincial Archives of Alberta photographed the first flight covers used in the book, and produced many other prints at short notice.

Charlene Dobmeier, managing editor at Fifth House Publishers, shared her expertise at great length as she guided a novice through the publication process. Her time was my time, and I am grateful for her unwavering commitment to this book. Donald Ward edited the manuscript with patience and care.

While these contributions have improved the book, any errors that remain are mine.

This book is dedicated to Alberta's sky riders from yesterday, today, and tomorrow.

A fair in Wetaskiwin: many hands help to hold a balloon down until it is fully inflated. Courtesy Provincial Archives of Alberta, A4871

Gas Bags and Bird Men

Excitement rippled through Edmonton in June 1906. Alberta's first provincial fair was to be held the following month, and a balloon artist had been booked as part of the entertainment. Professor R. Cross was to perform daily ascensions and parachute drops.

Anticipation grew as the exhibition drew nearer. The Canadian Pacific and Canadian Northern railways advertised special excursion fares into Edmonton from stops along their lines. City merchants decided to close their shops every afternoon of the fair. On July 2, opening day, the skies were clear and, according to the *Edmonton Journal*, "the whole city gave itself up to merrymaking."

Just after seven in the evening, the balloon was ready to go up. Several men from the crowd had been helping to hold it on the ground while it was filled with gas. On a given signal, they let go, and the balloon "darted like an arrow up into the air." Professor Cross was on a trapeze below the balloon. He was waiting until the craft had risen a suitable distance before tripping the rope that released his parachute.

But something was wrong. The great balloon began to sink. Cross tugged on the trip rope, but he couldn't release his parachute. The balloon continued its untimely descent, eventually falling near McKay School. Cross landed unceremoniously on the roof of a nearby house while the balloon described a lazy circle around the school walls.

The problem, the *Edmonton Journal* later reported, was that the rope was too thick. It would be replaced, and the ascensions and parachute drops would continue on the remaining days of the exhibition. Further ascensions, however, do not appear to have taken place. The wind changed direction, threatening to dump Cross into the North Saskatchewan River and destroy his balloon on the heavily treed river bank. The professor declined to perform under such conditions. Fortunately, he was not the only daredevil at the fair. Patsy the High Diving Dog and Mademoiselle La Tena—who, it was said, did a marvellous performance on a revolving globe—had also been featured in the grandstand show.

At the Calgary fair the following week, a balloon ascension and parachute drop by a certain Professor Williams went a bit more smoothly. As his balloon ascended, he did acrobatics on the trapeze dangling below. When his parachute released cleanly, he floated gracefully to earth. Unfortunately, as the *Calgary Herald* reported, he "dropped into the Elbow River below the cemetery bridge." Ballooning, apparently, was not for the faint of heart.

The possibility of flight had intrigued humankind for centuries. Roger Bacon (1214–1292) suggested filling copper globes with "ethereal air" that would lift them into the sky. Others suggested fire. Edmond Rostand, the creator of Cyrano de Bergerac, wrote a story featuring dew-filled globes, explaining that the daily disappearance of dew indicated that it rose to meet the sun. Leonardo da Vinci (1452–1519) applied his genius to the problem and came up with detailed conceptions of parachutes, aircraft, gliders, and mechanical devices called ornithopters that were harnessed to the body and powered by the arms and legs of the person it was attached to. It wasn't until the 1600s that a mathematician proved that human beings did not have the muscle structure necessary to support themselves in flight. Even so, optimistic "tower jumpers" continued to leap into space strapped to winged contraptions, only to crash to the ground below.

During the 1700s, hydrogen was isolated as an element, and this lighter-than-air gas was pumped into balloons. Hot air also became popular as a means of ascending into the sky. By 1800, performers were parachuting from their balloons, and they became crowd-pleasing acts at fairs and exhibitions throughout Europe.

Inventors were increasingly interested in the possibility of heavier-than-air mechanical flight. Floating about in balloons was all very well, but steering was difficult, if not impossible, and the operator was at the mercy of the wind. George Cayley (1773–1857), who founded the science of aerodynamics, identified the problem of mechanical flight: you needed to "make a surface support a given weight by the application of power to the resistance of air." Easy to say, perhaps, but much harder to achieve.

Experiments proceeded with gliders and steam-powered machines. Small motors were added to balloons, giving them both power and directional control. No scientist seriously doubted that mechanical flight was possible. It was only a matter of time before it was achieved.

In the middle of the nineteenth century, William Henson designed an "aerial steam carriage" which, with wings, fuselage, and a tail, foreshadowed the modern airplane. By the end of the century inventors in Europe, Britain, Russia, Australia, and America were all working on heavier-than-air flying machines. Societies and journals devoted solely to flight were popular. In 1868, London's Crystal Palace hosted the first aeronautical exhibition, and eager crowds came to see the fantastic craft on display. Gliders were travelling further as pilots learned to control their machines in the air. It was the development of the internal combustion engine that finally solved the problem of a power source light enough to permit a craft to become airborne, but still able to produce enough power to propel it through the air.

By 1900 all the pieces were in place. Inventors had a better understanding of aerodynamics, they were learning the importance of controlling their craft through

Fair-goers in Edmonton watch a balloon ascension. Balloon artists became crowd favourites at fairs throughout Alberta. Courtesy Provincial Archives of Alberta: Ernest Brown Collection, B5802

wing and rudder movement, and they were developing smaller, more powerful engines.

Wilbur and Orville Wright were the first to pilot the idea of heavier-than-air flight from theory to reality. The two brothers from Dayton, Ohio, had been interested in aviation since childhood, when they had conducted experiments with kites. They understood that the problems of balance and control in the air, of the relationship of wing shape to lift, and of the proper power source would have to be solved before practical mechanical flight was possible. The brothers experimented with manned gliders, and in hundreds of flights they tested their theories about wing stability and control in the air. They discovered that wings did not have to work in unison—that, in fact, a flying machine would remain stable only if the pilot could raise and lower the wing edges independently. They built a wind tunnel and experimented with wing shapes, discovering that long, narrow wings provided better lift than short, broad ones. Working together, they discovered that one movable rudder at the back of the craft (instead of several "fins"), when linked to wires controlling the angle of the wings, overcame the problems of air resistance that sometimes prevented the machine from holding a turn. They examined the shape and function of propellers with the same meticulous attention.

Then on 17 December 1903, with Wilbur as pilot, the Wright brothers flew their creation for the first time. It remained airborne for twelve seconds. Alternating as pilots, they took the machine up for three more flights. These flights were, in Orville's words, "the first in the history of the world in which a machine carrying a man had raised itself by its own power into the air in full flight, had sailed forward without

The Underwood brothers drag their elliptical kite down the main street of Stettler in 1907. The base of the kite rests on a stoneboat while one of the brothers steadies the frame from the back of the wagon. Courtesy Glenbow Archives, Calgary, NA–463–30

reduction of speed, and had finally landed at a point as high as that from which it had started."

Hopeful aviators around the world were still building gliders, experimenting with wings and small motors, and using kites to discern the workings of the wind. Alberta had its share of experimenters as well.

In 1909, two Calgarians tried unsuccessfully to launch their kite-like creation from the Bowness Trail with the aid of a single-cylinder motorcycle engine. J. Earle Young and Alf Lauder, twelve and fifteen years old respectively, had caught the flying bug early, and they did manage eventually to get their craft airborne by towing it behind a Buick.

In 1907 the Underwood brothers—Elmer, George, and John—brought the novelty of flight to Stettler. Their "flying wing" was elliptical, constructed of long strips of fir. Wire spokes were attached to the wings from a central hub that contained a platform for the pilot and the engine. Beneath the platform were two motorcycle wheels, while bicycle wheels on each wing helped the craft move on the ground. A rudder and stabilizing framework provided control. Canvas covered the wings. The frail but elegant machine was on display at the Stettler exhibition that July, to the delight of fair-goers.

The Underwoods did not have a motor, and had to test their machine as a kite. They tethered it to a fencepost on their farm and launched it into the air, using bags of wheat to simulate the weight of a pilot. John convinced his brothers to let him take the place of the wheat, and for about fifteen minutes he sailed above the ground, safely tied to the fencepost.

The next spring they linked a motorcycle engine to a bamboo and canvas propeller and prepared an airfield by packing down part of a field. The seven-horsepower motor easily propelled the machine along the ground, but it was not powerful enough to get it airborne. Unable to find a larger engine they could afford, they abandoned the idea of powered flight, and continued experimenting with tethered kite flights. A windy day brought their ambitions to an end as the

What do we call them?

The language of early aviation was colourful as inventors strove to name each new development or device. Balloon and gas bag were apt; so was dirigible, which means "capable of being guided." But what about those winged contraptions that flew through the air? Airships and flying machines, mechanical birds, aviators and bird men were popular names in the press for the flyers and the machines they flew. Aeroplane eventually became the most common term for these heavier-than-air craft. It was a combination of *aero*, which means air, and *plane,* which means a flat or level surface and referred to the wings.

giant kite smashed on the ground. The brothers chose not to repair it.

The Underwoods' kite flights, while demonstrating a measure of ingenuity and no doubt entertaining the neighbours, were not keeping up with developments in the world of aviation. Visitors to the Calgary fair in 1908 had been treated to the sight of a dirigible floating serenely over the city, a spectacle the *Edmonton Journal* described as "marvellous and thrilling," demonstrating "the wonderful strides the science of aerial navigation is making." The airship was owned by an American named J. Strobel and piloted by a Captain Jack Dallas. Dallas was accompanied by a group of assistants who dealt with the paraphernalia required for aerial navigation.

As described by the *Calgary Herald*, the airship was 60 feet [18 metres] long, contained 600 yards [550 metres] of Japanese silk, and held some 8,000 cubic feet [226 cubic metres] of hydrogen. The frame was crafted of Oregon spruce, the net of French linen. The propeller, made of oak and tin, made 350 revolutions a minute. A ten-horsepower, four-cylinder gasoline engine provided the power, with a large rudder at the rear for steering. Dallas, the *Herald* noted, had flown dirigibles all over America, but the article concluded on an ominous note, reporting that the captain had had several "hair breadth escapes from death."

Dallas awed the Calgary crowd with his ability to control the airship, moving it in all directions regardless of wind direction. Then on 4 July the great bag exploded and burned as a violent windstorm swept through southern Alberta. One man at the exhibition was killed and several others injured when tents and booths collapsed and poles and debris blew about. The airship was tethered at the time, but the wind seems to have forced the bag against one of the poles that anchored it to the ground. It rubbed against the guy wires that steadied it, and the resultant friction must have

Strobel's dirigible floats high above Calgary in 1908. Captain Jack Dallas works the airship from the gondola beneath the balloon. Courtesy Glenbow Archives, Calgary, NA-423-5

built up enough heat to ignite the hydrogen. According to the *Calgary Herald*, a pole fell on the airship, causing it to burst. The whole thing then exploded in flames. Dallas and an assistant, Bert Hall, had been filling the bag with hydrogen, a highly explosive gas, when the winds hit. Dallas received burns to his face and hands; Hall was burnt on the arm. The dirigible was completely destroyed.

It was doubtful that Captain Dallas would be able to make his next public appearance, which was scheduled for the Brandon fair, but he said he hoped to resume his engagements in a couple of weeks. When he left Calgary, he told the press that he was heading for Fargo, North Dakota, where he would perform later in July. His face was quite sore, he said, but he expected the burnt skin to peel off soon and he would be able to smile again.

While balloons and dirigibles continued to attract crowds across Canada, Alberta newspapers kept up with developments around the world. When an airship in Paris ascended 1,100 metres, they reported it, and Albertans also knew of Count von Zeppelin's impressive journeys by dirigible across Germany. An editorial in the *Calgary Herald* in August 1909 remarked that "The schedules of airship arrivals are already beginning to resemble CPR train bulletins. Yesterday at Berlin the notice posted up was 'Zeppelin No 3, 12 hours late.'"

Clifford T. Jones, an enterprising Calgary barrister, wanted to bring long-distance dirigible travel to Alberta. Jones applied to the Commissioner of Parks for permission to lease land on Cascade Mountain for an airship station. It was perhaps a bit premature to be setting up a network like this, he admitted to the *Herald*, but he insisted that there would soon be a need for such facilities.

Jones's scheme never got off the ground. Balloons and dirigibles were fine, but it was airplanes people wanted to see. Albertans knew that distance records were being smashed in Europe. Louis Bleriot, a Frenchman, had crossed the English Channel in a monoplane of his own design in July 1909. Wilbur Wright had flown in London in front of the British prime minister. In Canada, the Aerial Experiment Association had been formed in Nova Scotia in 1907 at the suggestion of Mabel Bell. Mrs Bell had offered to fund aerial experiments to be undertaken by her husband, Dr Alexander Graham Bell, and three young colleagues: John A. D. McCurdy, a family friend and recent graduate in engineering of the University of Toronto; Frederick W. Baldwin, an engineering friend of McCurdy's; and Glenn H. Curtiss, a young American inventor and expert on gasoline engines.

The group continued Bell's experiments with kites, then progressed to powered flight, moving their operations to Curtiss's workshop in Hammondsport, New York. By the spring of 1908 the association had an airplane in the air. Several more were built, each incorporating new features which had been carefully thought out and tested. Bell wanted one of the planes to fly in Canada, and shipped the Silver Dart to Nova Scotia in January 1909. Near the tiny community of Baddeck, Canada's first manned, powered flight in a heavier-than-air machine took place. On 23 February 1909, with McCurdy at the controls, the plane lifted into the air. Less than a kilometre later it settled onto the ice of Bras d'Or Lake.

Perhaps it was the seeming simplicity of early airplanes that encouraged inventors and amateurs to build them. A simple frame, some motorcycle wheels, a light motor, a propeller, and some canvas, and you were ready to go. To Reginald Hunt, a carpenter in Edmonton, it didn't look too difficult. He built a motor in his spare time, and designed a propeller that was, according to a newspaper article, based on "the fans that keep flies from sleeping in restaurants." On Labour Day, 1909, Hunt lifted off and stayed airborne for some thirty-five minutes. The following year, he flew at the exhibition, but crashed into a fence. This embarrassing mishap, coupled with his inability to get financial backing for a proposed plane factory, seems to have brought his career in aviation to a halt.

William Wallace Gibson came to Alberta in August 1911 solely to further his aviation experiments. He had found his work severely hampered by conditions in British Columbia. The clear blue skies and bone-dry weather of Alberta, not to mention the absence of trees on a friend's ranch near Calgary, were exactly what the inventor was looking for. Gibson had been experimenting with kites and then with planes for almost thirty years, starting as a boy in Saskatchewan. The air-

This photograph shows the narrow wings or "planes" Gibson used as the basis of his flying machine. Courtesy Glenbow Archives, Calgary, NA-463-35

craft he brought to Alberta, the Multiplane, was his latest design.

The *Calgary Herald* got hold of the story and splashed it across the front page: "Aviator Is Flying Near Calgary Daily; Has Unique Machine Called Multiplane." The machine was unlike the monoplanes and biplanes with which most people were familiar. The Multiplane got its name from its many narrow wings of laminated spruce. The *Herald* compared Gibson's airplane to a hay rack, or a huge bird cage. Gibson had opted for multiple small planes, or wings, rather than one or two large ones because he wanted every part to provide lift. His craft had a total lifting surface of thirty square metres, and the entering edge of the planes totalled 104 metres. It was the entering edge of each plane that did most of the lifting, Gibson explained; consequently, the machine had great elevating strength and stability in the air. It weighed 318 kilograms and had a sixty horsepower engine. Gibson maintained he could start the machine and send it into the air by itself. He simply set the carburettor so the motor would work for one minute, and when it stopped, the plane floated back to earth.

Gibson loved to talk about his machine. He told the *Herald* that he had constructed the engine from patterns he had made himself, and from "forgings milled and turned in his own machine shop." He had to use three carburettors to get enough air to mix with the gas to allow combustion in the engine, which he tested by running it for hours at a time.

After several successful short hops, it was time to try a longer flight. On 11 August 1911, with Alex Jaap, one of Gibson's assistants, as pilot, the Multiplane took to the air. In front of a small crowd of excited onlookers, Jaap flew about one and a half kilometres, climbing approximately thirty metres into the air. All went well until he cut the engine and began to glide down for a landing. From fifteen metres up, he saw badger holes honeycombing the landscape below. He tried to restart the engine, but it would not cooperate, and the plane continued its descent. Jaap headed for a slough. The landing was certainly softer in the mud, but the abruptness of the stop as the wheels bogged down tore the plane apart.

Jaap emerged with bruises, and his picture with the wrecked plane appeared in the paper. Gibson, ever the optimist, pointed out that another feature of his Multiplane was that the rear placement of the engine ensured that it would not crush an unfortunate aviator in a crash. Both Jaap and Gibson claimed to be "delighted" with the flight. Gibson announced that he would rebuild the machine with steel planes, giving it more strength. Unfortunately, his finances were exhausted, and he never did return to the project.

In March 1913, the *Herald* reported that a "Calgary man had built a biplane of Curtiss type" and that he planned to start an aviation school in the city. Joseph Simmer had spent two months at an air school in Chicago, studying "all aspects of aviation," and was planning to give exhibition flights in Calgary later in the spring. He did fly successfully near Calgary, but his school never got off the ground. He sold the plane to a man in Coronation, where it crashed on Victoria Day, 1914.

But perhaps no story captures the spirit of the times better than that of Tom Blakely and Frank Ellis. Blakely, manager of the Calgary office of a Winnipeg real estate firm, had acquired the remains of a Curtiss biplane that had been wrecked

Joseph Simmer, right, shows off his biplane. Courtesy Glenbow Archives, Calgary, NA-1239-1

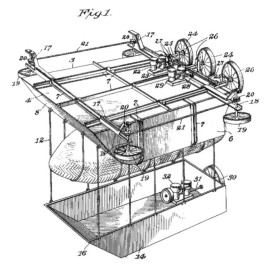

A patent drawing submitted by Newbury for his flying machine. Courtesy Glenbow Archives, Calgary, NA-2995-1

Frank Ellis and Tom Blakely get the "West Wind" ready for a flight near Calgary. Courtesy Glenbow Archives, Calgary, NA-2324-4

during exhibition flying in Saskatchewan. He placed an ad in a Calgary newspaper for "someone interested in aeroplanes" to help him rebuild the machine. Ellis, an enthusiast since childhood, answered the advertisement.

Blakely and Ellis carefully repaired the wings and undercarriage, fashioning new beams, struts, ribs, and spars from fir and spruce. Metal parts were repaired or replaced as necessary. The motor was overhauled, wheels were attached. Finally they took the machine out to a field they had chosen as their testing ground, and for the next few days practised taxiing, toyed with the motor, and repaired the frequent tire blow-outs. By the end of June 1914, they had succeeded in taking off, and for the next month they made many hops around the field.

They slept in a tent beside the plane so they could be up at the crack of dawn to fly in the still morning air. They hauled water in buckets from the river a kilometre or so away. Ellis was fired from his job three times for tardiness, but he was always hired back, and the summers of 1914 and 1915 were magical for the two young aviators as they lived the age-old dream of flight.

It was Alberta's weather that put an end to this particular dream. The sun had weakened the canvas of the wings, and one day a seam tore as Ellis was landing. That damage could have been fixed, but a subsequent windstorm tore the fragile craft from its stakes and tumbled it across the prairie. Blakely and Ellis salvaged the engine and the propeller, and left the rest where it lay.

No doubt there were other hopeful aviators building airplanes in barns or behind sheds, testing their theories on calm summer days, trying to coax their machines into the air. In 1911, Abel Newbury of Vermilion patented a flying machine which, made up of a gas bag between two propeller-powered platforms, looked like a dirigible sandwich. Earl Kelsey of Bawlf took out a patent the next year, but his plane was constructed along the more conventional lines of the mono- and biplanes of the day. John R. Hendrickson of Edmonton, on the other hand, developed a flying machine that featured several mysteriously placed wheels, one spinning over the top of the plane and another in the tail section.

Gradually, though, the airplane was being demystified. Alberta newspapers regularly covered aviation events around the world, often with photographs, and always with descriptions of the stunts performed. When an aviator plunged to his death, that too was reported. Most new developments in flying found their way into Alberta newspapers, including astonishing reports of new altitude, distance, or endurance records. But no practical use for the airplane had yet become apparent. Most of them couldn't carry more than one or two passengers, and none could be weighed down with freight. Their possible military uses were the subject of ongoing debate, but in the years before World War I the airplane remained largely a marvellous toy.

Crowds the world over enjoyed flying exhibitions, a fact that aviators, airplane manufacturers, and exhibition promoters picked up on quickly. By 1911 stunting exhibitions were popular in the United States, and many aviators included western Canada in their schedules. Most were with an exhibition company. One of the biggest was Glenn Curtiss's, run out of his flight training school in San Diego. Advance men travelled across the country booking performances. Local sponsors had to guarantee the aviator's fee if gate receipts didn't cover it. Planes were crated and shipped by rail from one destination to the next. A team of mechanics travelled with each flyer to assemble his plane and repair any damage it might suffer in flight.

In early April 1911, Edmonton's exhibition committee announced that it had booked Bob St Henry, known as "Lucky Bob" because of his several escapes from aerial disaster, for an exhibition on 28 April. An aviator named Hugh Robinson would also perform. The lead-up to the event included photographs of Robinson and St Henry posing with their machines and flying them at various locations. In an interview with St Henry, the flyer revealed that he kept his weight down to about fifty-seven kilograms by gymnasium work, cross-country runs, and careful eating. He didn't eat much on the day of a flight, he said, as the stunts tended to cause nausea, and he liked to keep his mind clear and ready for quick thinking. The interviewer remarked that St Henry looked more like a bank clerk than an aerial daredevil.

Finally, all was in readiness. The air and humidity had been tested and Edmonton declared a fine site for a flight. The field at the exhibition grounds had been found to be smooth and long enough to permit the machine to get airborne. The advertising had been done, calling the exhibition "A sensational and scientific exhibition of flying in the Curtiss Biplane, fastest in the world." The committee, in an attempt to make the exhibition sound educational, had announced that children would be admitted at half price so they would not miss "the scientific demonstration of the possibilities in the future of heavier than air machines." Regular admission for the show was $1—a bargain compared to the $2 many Edmontonians paid that same week to see "England's most Famous Singing Comedienne, Vesta Victoria, and her company of 20 international funmakers."

On 28 April, Robinson took to the air. Newspaper descriptions of the flight, not surprisingly, were filled with images from the bird kingdom. Robinson was a "bird-man" whose plane rose "like a bird of prey swooping up with a choice tid bit." Another writer likened his skill to "the ease and dexterity of an eagle." The next day Robinson and St Henry both made successful flights.

The exhibition ended on a slightly sour note. A real estate agent had thought he had gotten Robinson to agree to fly across the North Saskatchewan River and land on a plot of land in a new subdivision in Strathcona. People crowded the river banks on Sunday afternoon, 30 April, and searched the sky. But no airplane appeared. Disappointed, they headed home to wait for the next day's newspapers to tell them what had gone wrong.

While the crowds had been waiting, the men and their plane were already on a train heading out of the city. Several reasons were given as to why a flight across the river had not been made: they had to make connections to get to their next exhibition locations; they were afraid they might have been in contravention of the *Lord's Day Act* if they flew on a Sunday. There was another clue in the newspaper report: the aviators, it said, had left the city "not much impressed with the aviation enthusiasm shown by Edmonton." Perhaps the crowds had been too small for their liking, and they felt Edmontonians did not deserve another look at their flying prowess.

Still, Robinson and St Henry led off an exciting few months of aviation activity in Alberta. In early July Howard Le Van, flying a plane owned by the American J. Strobel, was at the Calgary fair. On opening day, however, he couldn't get the Golden Flyer into the air because of ground conditions. Mud clogged the wheels, and he couldn't get up sufficient speed for takeoff. He made several more attempts during the fair, and finally got airborne, only to crash into a fence. The wheels were broken and some of the silk on the wings was torn. After repairs, Le Van was able to get airborne again the next day. This time he flew for about five minutes, but an encounter with a gopher hole on landing damaged the plane further. There was no flying on the final day of the fair.

The newspapers declared the aviation portion of the exhibition a disappointment. The airplane was supposed to have made two flights a day. Instead, it spent most of its time in a tent where interested people could pay 10¢ to go in and have a look. Many did, apparently, as Le Van collected "considerable money," according to the *Calgary Herald*.

The citizens of Lethbridge were certainly not disappointed with the flying exhibition given there by Eugene Ely. Ely was a well-known American flyer whose exploits were widely reported. Most recently he had landed his plane on the American battleship Pennsylvania while it was moored in San Francisco harbour. On 3 July a contract was signed with Ely's manager, who declared Lethbridge's "climatic conditions and altitude ideal and the general conditions most satisfactory." The Lethbridge

Interested onlookers accompany Hugh Robinson and his plane to the centre of the exhibition track in Edmonton, 1911. Courtesy Glenbow Archives, Calgary, NA-1258-13

committee made sure that everything would be ready for the exhibition on 14 July. Special trains would be run from the city to Henderson Park for the event. As in Edmonton, the educational nature of Ely's exhibition was stressed.

By 13 July the "precious machine" had arrived in the city, but not without drama. Customs regulations had no provisions for airplanes, and local officials had to wire Ottawa for instructions. As long as the plane was going back to the United States after the exhibition, came the reply, it would be allowed into Canada. With that cleared up, the plane was ready to be assembled.

Lethbridge was in a holiday mood. Many businesses closed, although one advertised "A Flight of Shoes Not Airships at the Foot Toggery." On the day of the exhibition, people thronged to the park by train, by carriage, by car, and on foot. Ely flew twice, each time thrilling the onlookers. Dipping in pretence of landing among the crowd, gliding down to earth, and flying out over the city, Ely gave a good demonstration of current flying techniques. According to newspaper accounts, his two flights totalled some nineteen minutes, and the crowd of 5,000 went "wild with enthusiasm."

As quickly as it had begun, it was over, and the daredevil was on his way to Seattle to do it all over again. The *Lethbridge Herald* published two pictures of Ely in the air, noting with civic pride that he was flying over the new main exhibition building, which showed up "splendidly" in the photo.

The event that closed the 1911 flying season was the most ambitious planned so far. Didier Masson, a daring French aviator, was to perform in Calgary in October and then attempt to fly to Edmonton. The event was sponsored by the *Calgary Herald*, whose aim was "to give the people of this city and province an opportunity to see aerial navigation in its highest and most modern form." The plane, the aviator,

Eugene Ely soars over the crowd at Henderson Park, Lethbridge. Courtesy City of Lethbridge Archives and Records Management, P19740030054

and the mechanics all arrived. The plane was assembled and tested in a field outside the city. Pictures of Masson and his plane piqued local curiosity, and reports of his successful trial flights kept expectations high.

Everything went wrong. On his first attempt to fly to the exhibition grounds, he managed to get into the air, but engine trouble forced him to descend. Then the engine seemed to recover. He tried to rise again, but saw he could not avoid some telephone wires and was forced to land. On his second attempt to take off, he ran into some baling wire that caught in his propeller. Splinters of wood flying off the blade tore the silk on the wings and damaged the tail. Further flying that day was clearly impossible, and a car was sent to the exhibition grounds with the news. The disgruntled spectators were issued "wind checks" good for Masson's next scheduled appearance.

The *Calgary Herald* explained to its readers how uncertain the whole aviation enterprise was. Flights depended not only on the condition of the machine, but also on the weather, and therefore could not be scheduled with certainty. The *Herald*, perhaps, was regretting its decision to hold the meet so late in the year. This was Alberta, after all. "If the wind will go down and the snow keep off we may hope soon to witness a flight," the paper explained. "If fate and the weather should be against us we must accept the result with equanimity. Rome was not built in a day. Air flight exhibitions cannot yet be organized to take place on schedule."

No flights would take place if conditions were not right, the *Herald* stressed. People had been killed trying to fly in inhospitable conditions, and that risk would not be taken here. Less than a month before, the paper went on, an aviator had been killed at Dayton, Ohio, when his fuel tank exploded in the air. The jeers and shouts of "coward" from the assembled crowd had forced him to fly a plane he knew was not safe.

On 20 October, Masson got his plane into the air and flew over the city before returning to his camp. The motor wasn't working correctly. Masson thought he knew what the problem was: inferior gasoline. He ripped the fuel line off the carburettor and stuck it right into the air chamber. When he started the motor up again, it

In the Air: Monoplanes and Biplanes in 1911

Capitalizing on the interest generated by Masson's planned exhibition, the *Calgary Herald* carried a two-part series on "Modern Aeroplanes" in October 1911. The articles described the various aeroplanes being used around the world, and included a number of sketches and diagrams. The Curtiss Biplane, the writer explained, was distinguished by its three-wheeled "landing chassis" and by its "balancing planes" positioned horizontally between the ends of its two main planes. Monoplanes, having only one large wing, or plane, were described as popular because of their resemblance to the birds "we are so used to seeing in the air."

worked perfectly, and he flew to Victoria Park. Unfortunately, work horses pastured at the park had not been removed, and they raced about in alarm as the plane circled overhead. Masson had to make several passes in order to get them out of the way so he could land.

When he finally made his first exhibition flight, it was spectacular. He circled Victoria Park, waving to the crowd, then flew low over the city. He zoomed down toward the spectators as if he would land among them, then turned off the engine just in time and glided in for a soft landing on the field. By all reports the crowd went wild, and thronged around Masson and his plane. The police managed to keep order and the plane was not damaged. When Masson went up again, he returned to the same tumultuous appreciation. Many people implored him to take them up with him. He resisted all offers—even, the paper reported, "the charms and winning ways" of women.

Planning for his flight to Edmonton, Masson had driven the Edmonton Trail, picking out land marks so he could get on course immediately after takeoff. He had remarked on the cold in his short flights over Calgary, and he was planning to wear an Arctic cap that covered most of his face, high sheepskin boots, sheepskin leggings, and a suit of heavy tweed lined with leather on the long flight north. He would

Didier Masson ready for a flight. Courtesy Glenbow Archives, Calgary, NA-463-12

also be padding his clothing with newspaper, he said, for it made good insulation. Fur mitts completed the outfit. A special train would follow him on the ground, carrying his mechanics and their tools. A passenger coach had room for twenty people willing to pay $20 each for the honour.

The planned flight received a great deal of attention. Telegrams arrived from as far away as New York and Los Angeles, wishing Masson well on the journey. Letters with special letterhead commemorating the trip were ready to go up with the flyer, and the mayor of Red Deer extended an invitation to Masson to drop in for lunch as he flew north. Business people and civic promoters in Calgary and Edmonton hoped the event would result in economic benefits for their communities.

Snow and a brisk north wind made flying impossible for a couple of days. Then the weather cleared and Masson took to the air, but the wires holding his gasoline tank broke. The tank knocked him on the head and the wires got tangled in the propeller, breaking the blades. The plane landed safely, although Masson was dazed and unsteady. He only realized later how lucky he had been to escape death. The damaged propeller was the last one he had, and he made no more attempts to fly to Edmonton.

Masson's experience is a good indication of the general state of flying before World War I. Motors were unreliable and a constant source of worry. They weren't powerful enough to fly in any appreciable wind, and flyers had to wait for relatively calm days before venturing into the sky. Fuel quality varied widely. The planes, made of bamboo, canvas, and silk, were extremely susceptible to damage. Pilots sat in the open on small seats or, in some home-built planes, on kitchen chairs with the legs sawed off. As heavily padded as he was, Masson would probably not have survived the cold of a flight from Calgary to Edmonton that October.

Takeoff and landing were also not fully understood. Horses, baling wire, gopher holes, even people rushing onto the field were constant hazards. Not everyone realized that an airplane needed a long, smooth, clear stretch of ground to get up enough speed to become airborne, and another to coast to a stop after landing.

These factors combined to make flying a precarious business indeed. Three months after his exhibition in Lethbridge, Eugene Ely was killed at Macon, Georgia, when his aircraft plunged to the ground. Soon afterward, an editorial in the *Edmonton Bulletin* argued that aviation work should be concerned with improving its useful applications such as carrying mail, freight, and passengers. The writer condemned stunting as "sensationalist crazy acts."

Of course, the thrill-seeking continued. Like other infant technologies, aviation offered people the chance to do something different, and to do it first. And like any other event that combined skill with danger, the public couldn't get enough of it. Thousands watched aviators swoop and dive through the air, and then went home to read of new distance records, relay races featuring teams of flyers, and other daring manoeuvres performed somewhere else in the world. In Europe, aviators were vying for thousands of dollars in prize money, flying from Paris to London and back. Another contest offered $20,000 to the first flyer who completed a flight around Britain's perimeter. Closer to home, Hugh Robinson

This poster for the 1911 Calgary Industrial Exhibition showed a cowboy capturing a biplane in his lasso. Airplane stunt displays became star attractions at Alberta's bigger fairs. Courtesy Glenbow Archives, Calgary, NA–1473-7

was following the Mississippi to New Orleans. Aviation was the thrill of the day.

The 1912 Calgary exhibition featured Jimmy Ward, a stunt flyer with the Curtiss team. He was scheduled for two flights a day, and was reported to have signed a deal whereby he got no money if he did not fly. Fly he did, for thousands of cheering spectators. One flight the *Herald* described as "hair raising." Ward took his plane a long way out from the grandstand, then flew straight toward the race track at high speed, waiting until he was only about twelve metres above the ground before righting his plane and coasting in for a landing. When not in the air, the plane shared a tent with the elephants, where fair-goers could have a closer look at it.

No airplanes flew at Calgary's 1913 exhibition, but the popularity of flying had forced other entertainers to incorporate it into their acts. The fireworks display included "fireworks flying machines" called pyroplanes. They had no "guiding strings or attachments" yet they rose in the air, dipped low, and rose again as they lit up the sky.

General flying was curtailed during World War I by an Order-in-Council of the federal government that limited non-military flights. Frank Ellis and Tom Blakely, for example, had to get permission from the Mounties to continue flying their machine on the outskirts of Calgary. Flying at exhibitions did continue. Katherine Stinson, an American flyer who had thrilled audiences all over the world, brought her plane to Alberta in 1916, 1917, and 1918. Her daredevil stunts included looping-the-loop, doing spirals and figure-eights, flying upside down, and climbing straight into the air only to plunge straight down again, levelling off a few metres above the ground.

In 1916 Stinson—a "slip of a girl," as she was often described—appeared in Calgary and Edmonton. In Calgary she gave several performances, including a demonstration of night flying in which she attached smoke bombs to her plane and traced patterns against the sky. She flew to the Sarcee military camp outside the city and had dinner with the commanding officer, returning in time to give her evening performance before the grandstand.

Two directors of the Edmonton exhibition who had travelled to Calgary announced that Stinson's flying "exceeds any press notice ever written concerning her sensational flights." In Edmonton she danced an "airplane jig" in the sky on one day, and defied a strong wind to perform loop-the-loops on another. The wind was so strong, the *Edmonton Bulletin* reported, that when she was flying directly into it her plane seemed to stop.

Stinson returned to Calgary and Edmonton in 1917, but misfortune dogged her that year. In Calgary she was grounded by damaged engines. Once they were repaired she tried to race an automobile, but strong winds kept blowing her off course. Things did not improve in Edmonton. Dread gripped the crowd as the aviator lost control of the plane as she took off. She headed straight for the grandstand, then dove toward earth. She turned off the engine and managed to land. The aircraft was damaged as it skidded across the ground. She emerged unhurt but disappointed at her failure to give the crowd a good show.

When she did get airborne a couple of nights later, the crowd was treated to a display of aerial poetry. "Against the crimson sky the airwoman wheeled and banked like some gigantic bird," the *Edmonton Bulletin* wrote, "describing circles and loops with all her old freedom and certitude. A stream of smoke from the signalling apparatus trailed in a long plume in the aeroplane's wake."

Stinson also dropped a paper bomb into the trenches dug by the military to give fair-goers an extremely sanitized sense of conditions in France. She missed the trench slightly, but the sensation of wartime flying and its consequences were not lost on the crowd. Albertans had read of the aerial exploits of flyers in Europe, and they knew that a man from Calgary and another from Edmonton had trained at the aviation school Stinson ran with her brother and sister in Texas, and were now serving in Europe.

In 1918, unhappy with her performance of the previous year, Katherine Stinson arrived unexpectedly at the Calgary fair to make amends. She had sent her plane on the train with the race cars coming to the fair, and managed to keep her presence in the city a secret. On 1 July she flew from her camp at the golf and country club and landed in front of the grandstand, to the delight of the crowd. She flew every day of the exhibition.

There was bigger news to come: Katherine Stinson would carry mail from Calgary to Edmonton. It was only the second time mail had travelled by air in Canada. Three weeks before, the first airmail had been carried from Toronto to Montreal. Hastily written letters were dropped off at the post office. On 9 July, at 1:03 PM, wearing a heavy coat and carrying a horse shoe for good luck, Stinson took off from a hill outside the city.

Motor trouble forced her down fourteen and a half kilometres out, but mechanics rushed to the scene and by early evening she was once again in the air. All along the route the telegraph tapped out her progress to the waiting crowds in Edmonton. Passengers on the Calgary-to-Edmonton train "raced" her between Lacombe and Morningside, crowding the windows and doorways for a view. Shouts erupted from the grandstand crowd in Edmonton as a sharp-eyed spectator picked out her plane on the horizon. A large white arrow had been laid in the race track oval to mark her landing place. She circled a couple of times, then headed down to earth. The crowd rushed the oval. The police and fair officials were "swept out of the way like straws," reported the *Edmonton Bulletin*.

Stinson was escorted to the platform for a brief ceremony that included three rousing cheers. Then she was whisked away to the Macdonald Hotel for a rest. She had delivered 259 letters in a trip that had taken just two hours and five minutes.

The bird woman agreed to stay in Edmonton and perform during the exhibition. No one was disappointed, with the possible exception of Miss Louise Belmont. Miss Belmont was a celebrated balloonist travelling the fair circuit in western Canada. She gave marvellous demonstrations of ascensions and parachute jumps, opening three parachutes during the course of her descent. She used a hot-air balloon, and enjoyed a reputation as a daredevil. But there could be no doubt that she was no longer the star attraction at the exhibition.

Stinson not only gave daring aerial displays; she also drove a race car. On what she called a "slow track," she set a new one-mile [1.6 kilometre] record for women racers. The final night of the exhibition was declared Katherine Stinson Night, and she did not disappoint. For ten minutes she dipped and swooped above the grand-stand, and once again, fair-goers marvelled at how far aerial navigation had come.

Katherine Stinson gave several more performances in Alberta. At the Southern Alberta Amalgamated Fair and Stampede in Lethbridge, she made two flights, banking, diving, and stunting as well as flying over the city. The citizens of Red Deer had been waiting anxiously for their turn to see the famous flyer perform. At a meeting of the directors of the Red Deer fair, luncheon privileges were discussed, new classes were added to the swine judging, and two directors were authorized to travel to the Edmonton exhibition to try to make an arrangement with Stinson. They came back with a contract, and all everyone had to do was wait. They were not disappointed. Stinson performed well, and was hailed as the "great event of the fair." She flew for about fifteen minutes, and then made what she said was her best landing in Alberta. She was presented with a bouquet of sweet peas, and the crowd left still buzzing with excitement.

Camrose was not to be outdone, and had secured Katherine Stinson to fly at its fair as well. She flew from Red Deer to Camrose, following the CPR line to Wetaskiwin, and then heading over to Camrose. Despite a threatening storm, she ascended into the dark sky and delighted the fair-goers with a fine display of flying.

The few days of merrymaking at summer fairs and stampedes provided enjoyable distractions, but the grim realities of World War I were hard to ignore. Enlistment from the western provinces was high, and newspapers daily carried the horrors of trench warfare into every household. Fairs during the war years usually included a

Katherine Stinson captivated Albertans with her daredevil flying during the war years. This photograph shows her ready to make the first delivery of mail by air in western Canada. Courtesy Glenbow Archives, Calgary, NA-4350-1

military component: trophy displays, examples of work done by convalescing soldiers, Red Cross tag days, and other fund-raisers reminded people that fairs weren't just for fun. Organizers stressed that people would come away from one of Katherine Stinson's performances with a better understanding of wartime flying. Albertans knew of the exploits of wartime flyers across the Atlantic. Some of them were local boys. Men were being trained as flyers at a camp outside Toronto, while in the centre of that city Curtiss JN–4s were being built for the Royal Flying Corps. As flyers and as factory workers, Canadians were making significant contributions to the aerial war effort.

What would Alberta's skies look like after the war? Were Katherine Stinson's cross-country flights a portent of things to come? As 1918 drew to a close, peace stopped being just a hope, and Canadians everywhere started thinking about what life without war might be like.

"Speed . . . and a
Clear Highway in the Air"

On Tuesday, 29 April 1919, the House of Commons met at 3:00 PM in Ottawa. After dealing with an item concerning railways, canals, and telegraph lines, the Honourable A. K. Maclean rose to speak on behalf of the Minister of Naval Service. He had before him a draft of Bill 80, an act dealing with aeronautics in Canada.

Maclean argued that the government needed to set up a board to supervise the development of aviation in the country. The board would oversee all matters concerning aeronautics, including aerial navigation and routes, developing regulations, building aerodromes and air harbours, licensing pilots and aircraft, undertaking aeronautical research, and cooperating with the authorities of other countries. This seemed to Maclean the most practical method of dealing with what he called "this new subject." Other Members agreed, and the motion to introduce the bill was passed.

The legislation proceeded to second reading, where it met with more strenuous debate. What would be the salaries of board members? Would there be an effort to employ the Canadians who had rendered such splendid service in the skies over France? Could aeroplanes be used for locating schools of mackerel, or for rescuing shipwrecked passengers and sailors? Finally, would a number of aeroplanes be reserved for the exclusive use of government ministers? Maclean patiently answered all questions (the answer to the last one was "no") and piloted the bill through the House.

Its full title was *An Act to Authorize the Appointment of an Air Board for the Control of Aeronautics*, and its existence indicated how far aviation had entered into Canadian life. Only fifteen years earlier the new provincial government of Alberta had been struggling with legislation to regulate the use of automobiles. Now here was a machine that flew, and federal authorities were stepping in to regulate its use. The exploits of World War I pilots over France had certainly raised the profile of aeronautics in Canada. It had even been proposed that airplanes be used to patrol Canadian waters for U-boats. Aviation had a proven military purpose. Could it have an application in peacetime as well? Might it become useful in civilian life?

The Air Board immediately grasped the importance of the airplane to Canada's vast geographical mass. By using what the *High River Times* called "speed . . . and a clear highway in the air," the plane could make distances seem smaller and the constraints of time less onerous. The board's first act was to pass an order prohibiting dangerous flying. By mid-summer of 1919 the Flying Operations Branch, which looked after civil government flying, was up and running, as was the Certificate Branch, which dealt with the licensing of personnel, aircraft, and air harbours. The board was drafting legislation to regulate every aspect of aircraft operation in Canada, and with the passing of Air Regulations 1920, Canada moved into the air age.

Everyone Into the Air!

In the years immediately following World War I, many aeronautical activities got started in Alberta. Albertans seemed remarkably "air minded," a popular term that suggested an interest in and general support for aviation. To a large extent it was the legacy of the war that got them going and kept them going.

Canadian civil aviation got a big boost from Great Britain. To help the empire get started in aeronautics, Britain made war surplus planes and aviation equipment available to member nations. Canada received eighty planes as well as parts, wireless equipment, hangars, and various storage sheds. The planes spent some time at Toronto being overhauled before they were sent on to their final destinations, including clubs in Edmonton and Calgary.

Other planes were available, too. War surplus Curtiss JN-4s made in Canada or the United States could be purchased for around $2,500. Thousands of veterans from

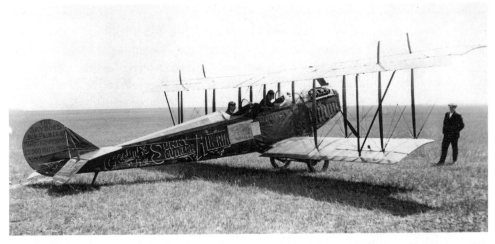

The "Skyboes," Jock Palmer and Harry Fitzsimmons, with their plane in 1922. "Welcome to Sunny Southern Alberta" is emblazoned on the plane's side, while the tail proclaims "22,000 miles [35,400 kilometres] in 1920." Courtesy Glenbow Archives, Calgary, NA-2002-2

every branch of the military were finding the transition back to civilian life difficult. Airmen realized that their flying skills could be transferred to the postwar world and were willing to give it a try. This core of trained and experienced flyers was a tremendous help in the development of postwar aviation in Canada.

The Lethbridge Aircraft Company, formed in 1920 by two veterans, is typical of early postwar ventures: a couple of men scraping some money together and calling one or two airplanes a company. H. H. Fitzsimmons and J. E. Palmer sold 100 shares at $100 each to finance their enterprise. To encourage investment, the company put

The Reynolds 1919 Sport Monoplane

For some people in Alberta after World War I, the thrill of the air meant building their own plane. Wetaskiwin's Edward A. (Ted) Reynolds was one of those determined flyers. Drawing on his mechanical background, Ted built a light monoplane in 1919. He used wood for the fuselage and wings, linen fabric for the covering, and aluminum engine cowlings. A Model T Ford car engine installed backwards, with the propeller instead of a flywheel attached to the crankshaft, provided the power.

Ted made several short flights in the little single seater monoplane during the next three years, but the first one on 6 September 1919 was perhaps the most exciting. His takeoff was down a hill, and he just cleared the fence at the bottom by a small margin.

The sport monoplane was restored by the Reynolds Aviation Museum in 1990 and is on display at the Reynolds-Alberta Museum in Wetaskiwin.

Ted Reynolds smiles proudly from the cockpit of his monoplane in 1919.
Courtesy Stanley G. Reynolds, Wetaskiwin, Alberta

out a prospectus describing its operation and the areas in which the men wanted to become involved. The list was ambitious: carrying passengers and goods, aerial advertising, exhibition flying, photography and survey work, instructing, selling and repairing airplanes—in short, anything to do with planes and flying.

The *Lethbridge Daily Herald* reported the arrival of the company's Curtiss JN- 4 in mid-May 1920, and proudly declared that Lethbridge was now on the "air map." By the end of May, eager citizens were lining up to be taken for rides. Public interest in the flyers was high, and newspaper accounts profiling Palmer the pilot—Fitzsimmons acted as business manager—stressed his impressive war record.

Captain Fred McCall of Calgary was another flyer who capitalized on his reputation as a war hero. Hired as a stunt pilot by the Canadian Fair Management Company in 1919, his first engagement was the Calgary fair. He flew Brigadier General H. F. McDonald to the grounds to open the fair. He executed an "Immelman turn" before landing, pointing the plane's nose up, then turning the plane over sideways and heading in the opposite direction. It was the general's first time in the air, and he claimed to have enjoyed every minute of it. With the general safely on the ground, McCall took off again and performed a dizzying array of stunts above the heads of the spectators.

Advance promotion for the exhibition stressed McCall's military record. The *Calgary Herald* carried a colourful description of how an aerial patrol he led had dispersed a German patrol. Dubbed "Calgary's Aerial Wizard," publicity pictures of McCall often showed him in flying scarf and goggles. His flare for the dramatic combined with his physical courage proved irresistible to crowds in southern Alberta.

McCall expanded the military theme during Calgary's fair by dropping a "bomb" on the front lawn of a prominent businessman. He circled Patrick Burns's house three times, then dropped to twenty metres as his passenger, Lieutenant Harry Payne, dropped a "cone-shaped leather case" on the lawn. The case contained an honourary membership for Burns in the Calgary Aero Club, and the whole event was described in the *Herald* as a "thrilling exhibition" of how "intrepid airmen" had bombed the enemy in France.

McCall did evening shows during the fair. Afternoon shows were put on by another veteran, Captain Wilfrid Reid "Wop" May of Edmonton. May had tangled with none other than Baron von Richthofen before a fellow pilot sent the Red Baron to the ground in flames. May's stunts were also thrilling, but more adulation was reserved for home-town boy McCall. Indeed, it was McCall who inadvertently provided the most exciting moments of the fair. When his plane lost power suddenly, he had to choose between landing on the crowded midway, on the track where a car race was in progress, or on top of the merry-go-round. He chose the merry-go-round. The pole and guy wires provided enough support, although the plane was badly damaged. Neither McCall nor the two small boys he was carrying as passengers were hurt. It was a heart-stopping moment. McCall's handling of the situation was described as a "wonderful exhibition of daring aviation, quick thinking, and pluck." It was no wonder, the *Herald* concluded, that he had been so successful in France.

The 1919 Red Deer fair was billed as the Victory Fair, and it also featured a war hero. Lieutenant George Gorman of Edmonton, the advertisements stressed, had "done his part" during the great conflict. His plane had been shot down behind German lines and he had been a prisoner of war for four months. In his bright blue plane with red trim, Gorman effortlessly performed dips and rolls and nose dives that stopped just short of the grandstand.

Gorman was flying for May Airplanes Ltd, a company founded by Wop May and his brother Court in the spring of 1919. They had signed an agreement with Edmonton to rent its Curtiss JN–4 Canuck biplane, called the "City of Edmonton," for $25 a month. As part of the agreement, May Airplanes agreed to promote flying in Edmonton and around the province, keep the plane in good condition and in proper storage, not let any unqualified pilot fly the plane, and not remove it from the province without the written permission of the city commissioners. Pete Derbyshire, who worked in the May family garage and had worked on airplanes during the war, was hired as the company's mechanic. The company established an aerodrome on a farm on the St Albert Trail.

With the formation of May Airplanes, the *Edmonton Journal* predicted the city would become a centre of air activity. According to the *Journal*, the company had

Fred McCall's plane rests atop the merry-go-round where he was forced to land after losing power during a flight. His masterful handling of a dangerous situation enhanced his reputation as a war hero and daring aviator. Courtesy Glenbow Archives, Calgary, NA–1111–13

EDMONTON JOURNAL

AIRPLANE EXHIBITION

Lieut. Geo. Gorman of Edmonton

has just completed the

First Air Delivery of Journals to Wetaskiwin

The Airplane is the Curtis Military Triplane "City of Edmonton," painted with the markings of the Royal Air Force as used on active service.

Saturday, June 7th
1919

George Gorman tipped his wings to Edmonton before starting for Wetaskiwin to deliver Edmonton Journals. On the way back, he dropped these flyers announcing the successful delivery. The aircraft is incorrectly identified as a triplane. Courtesy Stanley G. Reynolds, Wetaskiwin, Alberta

Spectators get a close look at the "City of Edmonton" as it takes off from the airfield at Sproule Farm to deliver newspapers to Wetaskiwin in June 1919. Courtesy Glenbow Archives, Calgary, NC-6-4326

contracted to fly the smaller country fairs, including Camrose, Red Deer, Lloydminster, North Battleford, and Prince Albert. The company was also available for police work, advertising, and deliveries. As Wop May was already under contract to McCall Aero Corporation for summer flying at some of the larger fairs, he and Court decided to bring in another pilot, George Gorman, to help with their schedule.

Although most people were familiar with airplanes, many had difficulty envisioning their usefulness in everyday life. May Airplanes, with the support of the *Edmonton Journal*, embarked on a couple of stunts guaranteed to raise the profile both of the company and of aviation, too. On 7 June, with Gorman at the controls and Derbyshire in the other seat, the "City of Edmonton" took off for Wetaskiwin, carrying two bundles of the *Edmonton Journal*. A crowd was waiting at the fairgrounds. The plane circled twice, dropping a bundle of papers each time. Then Gorman and Derbyshire winged their way back to Edmonton. Later that month, the *Journal* hired the company again for a similar excursion, flying to the Killam Sports Day with copies of the paper. This time the flyers landed, and eager spectators quickly bought all the papers.

May Airplanes had a busy summer. Along with flying at small fairs, Gorman flew at other shows including the Big Gap Round Up in Neutral Hills. The airplane performances were one of the many advertised attractions that included "Plenty of Hay and Water for Horses," and an "Automobile Repair Station on the Grounds." In Edmonton, May and Gorman gave a thrilling performance at the exhibition on opening day. They took off one after the other, "silhouetted against the golden sunset," the paper reported. They played tag in the air, and executed single and double loops. May performed a stunt known as the "Falling Leaf," in which the pilot pulls the plane's nose into the air so that the engine stalls, then lets the plane tumble through the sky. Just as the crowd fears it's no longer a stunt, he starts the engine and pulls out of free fall.

September 1919 gave the fledgling company more opportunities to showcase the possibilities of the airplane. In late August a policeman was killed in Edmonton. The search for the murderer soon expanded to Edson, and Wop May was engaged to fly a detective to that town to hunt for the suspect. An arrest was made soon afterward. Later that month, the Prince of Wales arrived in Edmonton as part of his postwar victory tour. May met the royal train as it neared the city, and escorted it into Edmonton. All on board, according to the *Journal*, were impressed with his expert handling of the plane.

Planes seemed to be everywhere. They flew over the Crowsnest Pass taking motion pictures of towns and scenery. They dropped advertising leaflets and coupons over Calgary and Lethbridge. On the 249th birthday of the Hudson's Bay Company, the newly established Webber Aero Service–accompanied part of the way by a puzzled hawk–flew over the Calgary store sending streamers floating down into the city.

Webber Aero was founded in March 1919 by George Webber, a Calgary garage owner. He had joined the Royal Air Force late in the war. On his return to Alberta, he started up his garage again and expanded into the air business. He purchased a Curtiss JN-4 that had been used for training in Canada, built a landing strip and a hangar, and hired Frank Donnelly as pilot. Donnelly made a few flights around Calgary before a leaking fuel line caused engine failure and the plane crashed on takeoff. No one was injured and the plane could have been repaired, but Webber seems not to have bothered. Webber Aero Service, however, does have the distinction of being the first to fly passengers around Calgary.

Passenger carrying as part of exhibition flying remained popular in the early 1920s. Fair organizers tried to put together the best package of attractions that they could, and airplanes were a big draw in the immediate postwar years. The Western Canadian Association of Exhibitions booked acts and travelling show companies for the fair circuits. Smaller associations booked for the smaller one-day rodeos and fairs that proliferated throughout Alberta during the summer and early fall.

Curiosity about airplanes, the thrill of a new experience, and the popular draw of aerial war heroes meant many people were willing to pay $15 or $20 for a ride, equivalent to a major furniture purchase. Fred McCall, Jock Palmer, George Gorman, Harry Fitzsimmons, and others flew throughout the province to fairs and rodeos to

From Foremost, where $10 got you a three-minute ride in Jock Palmer's plane, to the Crowsnest Pass where $5 bought you a flight around Sentinel Mountain: all over Alberta, barnstormers landed on ball fields, farmers' fields, and at the edges of small towns hoping to take advantage of the public's fascination with airplanes.
Courtesy Glenbow Archives, Calgary, NA-2604-39 and NA-3903-145

perform stunts and to take the local citizenry up into the clouds. Some of the towns visited by the McCall Aero Corporation in the summer of 1920 included Bassano, Brant, Brooks, Cereal, Drumheller, High River, Innisfail, Macleod, Medicine Hat, Munson, Olds, Okotoks, Rockyford, Stettler, and Taber. Gorman's destinations included Wainwright, St Paul de Métis, Athabasca, Kitscoty, and Colinton. If they didn't have any engagements booked they might perform "on spec," circling a town in hopes that a crowd would gather. The flyers would then land and see if anyone wanted to be taken up. If the prospects didn't look good, they took off to another town.

For many people in smaller centres, seeing a "barnstormer," as they were popularly called, was their first experience with airplanes. Barnstorming was often a novel experience for flyers as well, for they had to learn the tricks of their trade as they went along. Flying over enemy lines in France might have had its dangers, but so did landing on bumpy fields in gusty winds with eager children rushing to be the first to see the plane and pilot up close. Flyers and their mechanics, who usually travelled with them, learned to look for trees, the lee of a hill, dips in the ground, anything that would shelter the plane from the elements once they had landed. They learned to carry enough pegs and rope to tie the plane down. They learned to repair torn wings with linen, needles, and thread. Most of all, they learned that no one was ever fully prepared for a life of barnstorming through Alberta.

New experiences awaited the flyer at every turn. Harry Fitzsimmons published his experiences in a series of columns in the *Lethbridge Herald*, and later in a book. He wrote of birds smashing into propellers, of sleeping under the wings of his plane on the open prairie, and of friendly cows damaging the machine as they tried to

Many people took up aviation after the war, hoping to make a career of it. This plane belonged to Clyde Holbrook of Youngstown and Les McLeod of Hanna, who barnstormed around southern Alberta between 1919 and 1921. Here, just outside Hanna, a small crowd has gathered to have a look. Courtesy Glenbow Archives, Calgary, NA-3596-60

scratch their hides on the rudder and wing struts. Landing in rough fields where gopher holes and cattle wallows swallowed the wheels had disastrous results for the planes. Getting them repaired in the middle of nowhere could be a real challenge. Sometimes the local blacksmith was called on to make parts, but more often the pilot and his mechanic made the repairs themselves with whatever materials were at hand. Appearance counted for nothing; getting the plane operational was all that mattered. If it could not be repaired where it was, Fitzsimmons had it hauled by horse and wagon to the railway where it could be crated up and taken back to Lethbridge, where parts were ordered from the manufacturer and repairs done at the hangar.

Read Here Particulars

OF THE BIG

Calgary Exhibition

June 26th to July 3rd

1 9 2 0

CHANGING PLANES IN MID-AIR

Lieut. Ormer Locklear as he will appear in his Aeroplane Acrobatic Act at the Calgary Exhibition on First Four Days.

The program for the Calgary Exhibition of 1920 featured an illustration of just one of the daring stunts Locklear would perform.
Courtesy City of Edmonton Archives, Edmonton Exhibition Collection, MS 322, Class 9, s/c 3, file 9

Wherever they went, excited people ran out to meet them. Many wanted to be taken for a ride, few knew what to expect. One woman, after a landing that left the plane with one wing torn off and the propeller and some of the rigging broken, declared that everything had been lovely. She had assumed all landings were rough.

Most barnstormers realized early that loops and rolls and nose dives would not satisfy the crowds forever. Audiences were forever demanding new thrills, new ways for the aviators to cheat death. Consequently, local flyers were quite interested when, in the summer of 1920, Hollywood movie star Ormer Locklear brought wing-walking and other stunts to the Calgary and Edmonton exhibitions. Locklear specialized in changing planes in mid-air, and in scampering all over a plane while it was in flight. His latest movie, *The Great Air Robbery*, was playing in Calgary at the time, and he received a good deal of press coverage in the days leading up to the exhibition. According to the *Herald*, his performance lived up to all the advance publicity. Locklear was seen standing on the wings "in all positions," then leaping from one plane to a rope ladder dangling from another. The *Herald* did acknowledge that it was "doubt-ful everyone enjoys this act. It is a risky looking piece of business, and the upper-most thought in one's mind is whether he will make the grade safely."

In Edmonton it was the same story. "Locklear Leaps from Plane to Speeding Plane as Crowds Below Hold Breath in Wonder," the *Edmonton Journal* exclaimed. Locklear gave a "nerve-burning" performance as he stood on his head on the top wing, swung by his feet from the landing gear, and leapt from plane to plane without a rope ladder.

The novelty of Locklear's performances and the reaction of his audiences to his aerial tricks were not lost on local flyers. George Gorman and Pete Derbyshire got into the act quickly. Barely a week after Locklear's act in Edmonton, Derbyshire put on a show that, to the partisan local crowd, rivalled Locklear's. Derbyshire decided to parachute from an airplane. After a bit of practice, he and Gorman were ready. The parachute was tied to the undercarriage of the plane. When the plane had reached a certain height, Derbyshire climbed out of his seat onto the wing and hooked the parachute onto his harness. He jumped off the wing, and dangled below it by the rope that tied the parachute to the plane. He used something the *Journal* described as an "automatic knife" to cut the rope, then he "dropped like a bullet" for a hundred metres or so before his parachute opened and he drifted to the ground. Derbyshire wore a "life-saving coat" filled with a cork-like material that would keep him afloat if he happened to land in water.

Gorman and Derbyshire took their parachute act to the Red Deer fair, where the directors had agreed to pay the pair $350 for three flights and two jumps. When the show was over, the *Red Deer Advocate* reported that "the aeroplane work of Capt. Gorman was never more interesting and attractive, while the parachute drops came off perfectly and were a tense and pretty sight."

Other companies, including the Lethbridge Aircraft Company, added wing-walk-ing and other dangerous tricks to their repertoires. Fitzsimmons argued that it gave him the edge in securing fair contracts, as not all barnstormers could boast these stunts. It brought him a good dollar, too.

Fitzsimmons and Palmer practised their routine over a lake near Lethbridge. If Fitzsimmons got into trouble while he was balancing on a wing or clinging to the landing gear, Palmer could cruise low over the lake and Fitzsimmons could drop into the water. According to Fitzsimmons, though, the key to their performances was not water but air. They always made sure they did their act at a sufficient height so that if the airplane had any problems, Palmer would have enough room to glide down to earth. They eventually performed all across southern Alberta, but the first time was right over Lethbridge. According to the paper, thousands lined the streets, businesses closed, and street cars stopped as Fitzsimmons crawled all over the outside of his plane.

For most barnstormers, though, the routine didn't vary much: flying from town to town hoping for passengers, and fulfilling contracts for exhibition and stunt flying during the short Alberta summer. By 1921, though, that was starting to change. Economic conditions had worsened in the province. Grain prices were dropping after their wartime highs, and farmers who had bought land or mechanized their farms in response to wartime prices and demand found that they could no longer make their

Some businesses and charities used the airplane to promote products and causes just after World War I. In May 1920, Mrs Bulyea took to the skies with May Airplanes to publicize her fund-raising campaign for a new YWCA building in Edmonton. While in the air, she dropped copies of the Aerial News *that carried information about the need for the building. Here, Mrs Bulyea and pilot George Gorman pose by the plane in goggles and fur-lined helmets just before the flight.* Courtesy Glenbow Archives, NC-6-5262

payments. Small town fairs were cancelled; others suffered drops in attendance. Fewer places could afford the expensive aerial acts, and fewer people could afford the luxury of an airplane ride.

As early as 1920, McCall Aero Corporation had dropped its price from $15 to $10 for a fifteen-minute ride. Running a barnstorming business did take cash. The Lethbridge Aircraft Company estimated the cost of a 5.5 hour flying day at $24.80 for gasoline, $3.50 for oil, and $50 for wages for the pilot, manager, and mechanic. If

What happened to the "City of Edmonton"?

Unable to make a living solely from flying, May and Gorman returned the plane to Edmonton in 1924. Its wings were removed and it was stored in the horse barn at the exhibition grounds. Peter Lerbekmo from the Tofield area bought it in 1926 and restored the air frame to flying condition. In exchange for an interest in the plane, Ted Reynolds overhauled and installed an engine. He made several test flights in July 1928. In October he traded a 1927 Star automobile for Lerbekmo's remaining interest and became the sole owner of the "City of Edmonton." He flew the plane in the Wetaskiwin area for several years before putting it in storage. In 1980, staff at the Reynolds Aviation Museum decided to restore it, and today it is on display at the Reynolds-Alberta Museum in Wetaskiwin.

Ted Reynolds stands with the "City of Edmonton" at Coal Lake, Alberta, on 20 April 1929. His log book for that day reads, "Landed on slush ice, plane left till morning, ice will not hold her up." Courtesy Stanley G. Reynolds, Wetaskiwin, Alberta

depreciation on the plane were added in, it was another $6, and more was added if the pilot and mechanic wanted to eat, or spend the night somewhere more comfortable than a haystack. Resourcefulness could solve some of these problems; Fitzsimmons would scour the picnic grounds after performances hoping to find an abandoned hamper.

Airplanes aged, and most companies couldn't come up with the money to purchase replacements. Often the cost of repairs or parts (a propeller cost at least $80) after a crash landing was enough to drive an entrepreneur out of the business. Records from the May brothers' companies show how perilous an existence aviators led in Alberta in 1921. As of 11 April that year, May–Gorman Airplanes had taken in $15,794.75. The bulk of it came from shareholders ($3,141), passengers ($5,585), and fair contracts ($3,900). Expenditures totalled $15,713.43, including $2,378 for gasoline, $1,153 for insurance, $1,216 for parts, and $2,609 for the plane itself. After wages and other expenses such as advertising had been paid, the company had a total of $81.32 in the bank.

Still, they were heady days. For many Albertans the early 1920s was a time of excitement and novelty, of technological wonders, and the chance to meet a genuine war hero. The immense respect accorded veterans had helped fuel the enthusiasm of ordinary Albertans for the airplane. The papers had been full of letters from the front, of dispatches glorifying the courage and sacrifice of the soldiers. Few households emerged untouched after the slaughter, whether it had been a relative, a neighbour, or a church member they counted among the fallen. After the war, reintegrating

Making an adjustment to the engine before take-off in a Curtiss JN-4 Canuck biplane near Manyberries in southern Alberta. Courtesy Glenbow Archives, Calgary, NA–2020–5

veterans into civilian life occupied the public's mind. In considering applications for positions advertised by the Air Board, for example, preference was given to men who had served in the armed forces.

The Great War Veterans' Association had chapters in most communities, and helped keep public attention focused on the war and on veterans' issues. Cities and towns everywhere erected memorials to the dead. Fairs exhibited crafts and other work done by convalescing soldiers. War trophies were on display, too. The University of Alberta exhibited a German Fokker biplane that had been presented to it by the military. The physical and emotional legacies of World War I were very much a part of life in Alberta following the armistice, and they fuelled the popularity of stunting and barnstorming in the years immediately after the war.

In other areas, aviation was coming into its own; it was looking to the future rather than to the past. The airplane was beginning to be seen less as a large mechanical toy and more as a method of transportation that would have far-reaching implications in many aspects of modern life. No longer just a curiosity, the airplane was gaining legitimacy as an integral part of Canada's national transportation network.

Over Hill, Over Dale . . .

The Air Regulations passed by the Canadian parliament in late 1919 covered all aspects of operating aircraft, airships, and balloons. All aircraft had to be registered with the Air Board and have a certificate of airworthiness. Canadian registration marks began with a G followed by a dash and four capital letters beginning with C. The regulations required the registration mark be painted on the top and bottom of the wings and on the fuselage.

Air harbours were also required to have licenses. They had to use a "recognized method" of indicating wind direction, such as a conical streamer or a smudge fire. Each harbour was required to fly a coloured flag to indicate the direction planes were to take on their approach if they had to circle the air harbour before landing: a white flag indicated a clockwise circuit, a red flag counter-clockwise. Air harbours licensed for night landing needed green and red lights to indicate directions. Each harbour was required to keep a log book in which the registration marks of each plane that landed or took off, the pilot's name, and the time were recorded.

The regulations covered the issuing of certificates to air personnel including pilots, engineers, and navigators. They specified the lights required on aircraft, and the signals to be used to indicate a plane in distress or to tell a pilot that a landing was not possible at that time. There were rules for approaches and takeoffs around busy air harbours, and a specific prohibition against "Acrobatic alightings . . . at licensed aerodromes." Explosives could not be carried aboard aircraft, and aircraft had to fly over cities and towns at a sufficient height to allow them to glide to earth outside the settled area should the engine fail. Every plane had to carry a log book. Customs regulations were spelled out in detail, including instructions on how to unload an international plane that crashed in Canadian territory.

The Air Board wanted aviation to enjoy the confidence of Canadians everywhere. To this end, it planned an ambitious cross-Canada flight in the summer of 1920. The Certificates Branch arranged landing and refuelling facilities across the country. The first leg of the flight, from Halifax to Winnipeg, would be undertaken by the Operations Branch in a seaplane. The second leg, from Winnipeg to Vancouver, would be undertaken by the Canadian Air Force using land planes. The entire exercise was to take forty hours and cost $7,000.

The Air Board hoped the project would demonstrate that long-distance flights had commercial possibilities, and that they could be accomplished without strain on either the pilots or the machines. The board also felt the flight would assist recruitment for the Air Force, and create a general interest in aviation among the public. A departure date was set for late September.

After several false starts, equipment problems, and bothersome weather conditions, the flight finally got underway on 7 October. By the time the de Havilland 9A landed at Calgary, the head of the Air Board indicated that the lofty goals that had gotten the plane off the ground in the first place had been altered by the flight itself. The primary purpose of the flight, Lieutenant Colonel Tylee now said, was to gain experience. "We are not sacrificing the element of experience for speed or anything spectacular," he cautioned. The flight had already demonstrated that good ground support and knowledge of air currents were essential.

Snow, rain, and fog in the mountains delayed the departure from Calgary. On Wednesday, 13 October, the little plane took off over the Rockies. After several delays en route, it finally landed in Vancouver on Sunday. Total flying time had been forty-nine hours and seven minutes, spread over a little more than ten days.

Despite the setbacks, the Air Board stressed that long-distance flying during the day and night had been shown to be possible. It had also been shown that more aerodromes and wireless communication between planes and ground facilities were essential before these kinds of flights would be possible on a regular basis. "Wireless directional apparatus" to guide planes during night flying and in bad weather would also improve the potential of flight. The flight had underlined both the high and the low points of aviation in 1920: inadequate ground and navigation facilities, unreliable machines, and the significant role weather conditions played in successful flights contrasted with the skill of the pilots, their success with the most rudimentary landing facilities, and the sheer determination to make flight a viable mode of transportation.

The desire to go further and faster was common to many aviators after World War I. The horizon beckoned, and for the first time it seemed to be in reach. Using "speed . . . and a clear highway in the air," the airplane was beginning to tame the wild expanse of geography that had always affected transportation in Canada. If it was far away, aviators wanted to fly to it. If it was high, they wanted to fly over it. And so, on 7 August 1919, Captain Ernest Hoy stepped from his Curtiss JN-4, nicknamed "The Little Red Devil," onto the airfield in Lethbridge, having just flown solo across the Rockies from Vancouver.

He stayed less than an hour, as he was anxious to make his final destination, Calgary. On the way in he had flown low over Fort Macleod, circling the town and fairgrounds to give them a bit of a show for their annual fair. In Lethbridge he signed autographs, took on gas and oil, and picked up copies of the *Lethbridge Herald* to drop on the way to Calgary. His trip so far had been uneventful, he said, then climbed into his machine and into the sky.

He received a tumultuous welcome in Calgary. Over 5,000 people crowded the landing field. Members of the newly formed Calgary Aero Club surrounded the plane to protect it. Hoy praised the arrangements in Calgary, declaring that the lights and flares had been clearly visible in the night sky. He good-naturedly shook all the hands that were proffered, although the paper reported some embarrassment on his part in "dodging" the "effusive feminine congratulations" which were also offered. Tired and dusty, he finally escaped the crowd and headed for the Palliser Hotel for a bath and some sleep.

Hoy had used maps to help with navigation, although he found that even the best maps contained inaccuracies. The prevalence of ground mist meant that he had to rely on compass readings as well. He said he always kept a lookout for places to land when he flew, and there were many places—from sandbars to railway tracks to tree tops—to glide onto if the engine failed. He flew at about 1,700 metres, and said weather conditions had been quite good. He had left Vancouver before sunrise and watched the sun come up over the mountains. "Words cannot describe," he told

Captain Ernest Hoy coming in for a landing in Calgary. Courtesy Glenbow Archives, Calgary, NB-16-625

reporters, "the magnificent spectacle of the mountains seen under this light from a point high in the air."

Bad weather delayed his return trip until 11 August. He landed safely at Golden, British Columbia, but later taxied into some trees trying to avoid two boys who had run onto the field as he was trying to take off. With the airplane damaged, Hoy and his plane had to finish the trip back to Vancouver on the train.

American flyers were also demonstrating the possibilities of air travel, and General Billy Mitchell was a particularly vocal supporter of an expanded role for aviation in the military. One of his schemes involved flying four de Havilland planes to Alaska. The expedition took off from Mineola, New York, and began to wing its way cross country. On 27 July 1920, amid great excitement, the four planes landed at May-Gorman airfield in Edmonton. The flyers were fêted at dinners by civic authorities and local aviators. Bad weather delayed their departure for a couple of days, but then the foursome was on its way to Jasper, and eventually Nome, Alaska.

Closer to home, intrepid flyers were beginning to map out the future of aviation as a long distance carrier of people and goods, and as a method of transportation

Billy Mitchell's flyers drew a good crowd when they landed at the airfield on the St Albert Trail. Courtesy City of Edmonton Archives, EA-255-7

over inhospitable territory. In January 1920, the new Edmonton Aircraft Company announced that it had operating capital of $50,000 and two planes on order. It hoped to be running regular flights between Edmonton and Calgary by 1 April. A route to Peace River was also planned. The company was prepared to carry passengers, freight, and mail. Tentative ticket prices were set at $40 return to Calgary, $60 return to Peace River. Captain Keith Tailyour was both pilot and manager. Jock MacNeill, of Twin City Taxi, was also involved.

The inter-city flights did not get off the ground quite as planned, and Tailyour found himself performing stunts and carrying passengers at small fairs in central Alberta that spring. On 2 July he took a Mrs Jennings to Calgary in his open-cockpit Avro in the first passenger flight between the two cities. It took two and a half hours flying time, with a stop for fuel at Olds. Mrs Jennings, outfitted in motoring goggles, said it was a "perfectly wonderful trip" and she "thoroughly enjoyed it" despite the fact that there was rain at some points and they were heading into the wind.

Tailyour's furthest previous flight had been to Red Deer. His landing in Calgary, although without incident, does indicate the rudimentary state of navigation and ground facilities in 1920. "I could see you fellows waving your hats," he said to the small crowd that greeted their arrival, "so I had no difficulty in alighting."

McCall Aero Corporation in Calgary was also starting to carry passengers from point to point. In an address to the Kiwanis club, McCall had declared plane travel safer than riding in an automobile. People travelling on business were starting to recognize the advantage planes offered. The president of Premier Collieries made several trips to Drumheller from Calgary by plane, and declared it was the most efficient way to make the trip.

Perhaps the land north of Edmonton was the place where the "clear highway in the air" contrasted most sharply with conditions on the ground. Roads were few and anything but clear: they were more likely to be clogged with mud or under water. Expanses of muskeg, lakes, and forest stretched between small settlements. Although the railway had reached several points, and boats operated on the rivers during the summer, travel was still painstakingly slow. May–Gorman Airplanes, formed in early 1920 when the May brothers had increased their operating capital to $50,000 and brought George Gorman in as a partner, was the first serious aviation company in Alberta to turn its sights northward.

In late August 1920, Wop May took off for the Grande Prairie fair on a flight everyone knew was dangerous. Between Edmonton and his destination lay kilometres of empty space where a downed plane and its hapless aviators would meet almost certain death. May took Lieutenant Colonel G. W. McLeod, one of the directors of the company, with him. McLeod was an experienced surveyor who, it was hoped, could lead them back to civilization should they have to abandon their aircraft. With food stores for five days, they left the aerodrome on the St Albert Trail in the early evening of Thursday, 19 August. They stopped overnight at Whitecourt, leaving for Grande Prairie at 8:00 the next morning. Just two and a half hours later, their plane was spotted in the sky over Grande Prairie, and they landed to the delight of the fair-goers. For the next three days, people lined up for the chance to be taken flying. The airplane was the most popular attraction at the fair.

May spent six weeks performing stunts and carrying passengers at fairs and picnics in the Grande Prairie–Peace River area. But by late September, the cold weather was on its way and it was time to head south. This time he was accompanied by Pete Derbyshire, who had taken the train to Grande Prairie. Engine trouble forced a landing near Whitecourt. The two men headed for the Athabasca River, but soon became lost. Eventually they found themselves back at the plane. With no other options, they set to work repairing the engine. Remarkably, they made it hold together with pieces of wire and tape, and were able to fly out of their predicament. They flew to Sangudo, where Derbyshire decided enough was enough. He took the train back to Edmonton.

The trip illustrated just how perilous airplane travel still was. May had depended on compass readings, as there were no landmarks to guide him. Forced to land, and with no way to signal for help, it was ingenuity and sheer good luck that saved the two.

If An Airplane Swooped Down

"Have you figured out what you would do," the editor of the *High River Times* asked his readers in January 1921, " if you were driving along a road with a nervous horse and an aeroplane from the High River aerodrome suddenly swooped down and frightened the horse into a runaway that smashed your wagon and injured the animal and yourself?" Most readers had probably not considered the question. But perhaps some of them were struck by the contrast of the air age existing side by side with the horse and wagon right here in High River, Alberta.

The High River Aerodrome came courtesy of the Air Board. The board and other

---

Flying Impressions: A Grande Prairie Resident Takes a Spin with Wop May

Condensed from the *Grande Prairie Herald*, 31 August 1921

I approached the plane with a feeling of curiosity. True, I was curious, not to know how the machine would behave, but how I would behave. Below me on the roads I could see teams who looked like lazy ants, automobiles who were nice sized cockroaches slowly ambling along, individuals were only bits of white or color. But in an instant all changed—I was high enough to see a panorama unexcelled in my history. I saw our wonderful country from a low level, an intermediate and a high plane. How beautiful it is! If any one wants to know their heritage, get up under Captain Wop's care. You will be proud of yourself—but prouder of your country, incomparably more beautiful than words can depict it.

government departments understood how important air travel would become in those vast tracts of the country that were virtually inaccessible by land. The Forestry Branch of the federal Department of the Interior, for example, wanted to use airplanes to patrol forests and report fires, and the Air Board agreed to help. In August 1920 a station was set up at Morley with a canvas hangar and several de Havilland and Avro aircraft. Air patrols began in September and October. Pilots spotted fires from the air and relayed their location by wireless transmission to rangers below. It was much more efficient than the old system of lookouts; fire fighters could be rushed to the scene before a blaze got out of control. The aerial program worked so well that the department wanted to expand the patrols into other areas of the province.

Morley, however, was too close to the mountains, and the high winds made landings and takeoffs dangerous. The Air Board decided to move the base to High River, and in January 1921 the first men and equipment began to arrive. High River greeted the newcomers with interest, and chose not to speculate much about what would happen if an "aeroplane swooped down" on their horses and wagons. Early in February, the town sponsored a dance to welcome the men to their new home. The town hall was decorated with Chinese lanterns and miniature airships. The six-piece Allen Theatre Orchestra supplied the music. Dancing went on until after 3:00 AM after a midnight supper had fortified the revellers.

Construction continued in March and April. Hangars, a building for offices and storage, sheds, workshops, and a wireless tower all went up. The Air Board paid $340 to the American Lafrance Fire Engine Company for two twenty-gallon chemical engines, $1,017.96 to the S. F. Bowser Company for the 1,000-gallon underground gasoline storage tank and the curb pump, and $112 to one Charles Gustafson for clearing

Canvas hangars and de Havilland planes at the High River Aerodrome in 1922. The tall structure is the communications tower. Courtesy Glenbow Archives, Calgary, NA–1170-3

and harrowing the eleven and a quarter hectares of aerodrome surface. Other items, from carriage bolts to crescent wrenches to heaters, were purchased and installed to get the High River Aerodrome up and running.

The *High River Times* followed developments closely. "The aerodrome is fast becoming a popular spot for motorists from town and district," it reported in late April, "as the road leading to this station is daily dotted with motor cars bearing persons interested and eager to find all news available pertaining to this important

Assembling aircraft in a hangar in High River, 1921. These de Havillands were used for forest fire patrol. Because the planes had open cockpits, pilots wore sheepskin leggings and jackets with collars that came up over their heads, padded helmets, and padded gloves with a pair of silk gloves inside. Courtesy Glenbow Archives, Calgary, NA-2097-2

station and the work being done." A *Times* reporter was one of those clogging the road, and he was truly impressed by what he saw. The hangars "were cunningly spaced in exquisite scientific distance" and the planes were a "marvellous work of science." The airmen too were praised, and their war experience duly noted.

The airmen sought to return the favours shown them. In June, they sponsored a dance for the citizens of High River. Once again the town hall was gaily decorated, this time with bunting, flags, and lights. All dances were described in flying terms, and the dance floor was crowded as High River's own Star Orchestra filled the spring air with music. The *Times* declared it the best dance held in High River so far.

Flying patrols began in May, and the station proceeded to log 284 flights and over 700 hours in the air before operations shut down for the winter. Most of the flights patrolled the Crowsnest, Bow, and Clearwater forest reserves. There were refuelling stations at Eckville for flights to the more northern reserves, and at Pincher Creek for flights to the south. The planes also dropped educational leaflets at fairs and stampedes, warning about the dangers of fire and urging fire prevention.

One airplane went to Jasper and took the park superintendent for three days of flights. They collected information on lakes and mountains, and survey information that would later be useful for trail building. The airmen took part in other activities from High River. They performed a good deal of photographic work. They also helped search for the body of a man who had drowned in Lake Chestermere. Then in mid-summer, death touched the station more directly. A plane crash took the life of one of the pilots as he was taking off for a routine flight to the Clearwater forest reserve. Businesses in the town closed; as the *Times* reported, people "did not have the heart to work."

Still, it was clear that the High River station was there to stay. In 1922, wooden hangars on concrete foundations began to replace the canvas ones. Also that year, the pilots logged 267 flights for a total of 1,075 flying hours. The forestry patrols used de Havilland 4s that had been converted from double- to single-seaters. Every morning, one plane left High River going north, and another left going south, returning in the evening.

The Auditor General's report for 1922 shows a great deal of activity at the High River station. The work force included pilots, labourers, air riggers (who helped dismantle and repair aircraft), electricians, radio operators, photographers, navigators, and a cook. The base purchased four and a half kilograms of white linen thread for mending airplane wings, thousands of metres of tape and hoses, gasoline tanks, varnish, and everything else from carpenter's clamps to coal. Suppliers in High River and Calgary were the chief beneficiaries of the spending generated by the base.

The Air Board oversaw civil aviation in Canada until 1923. In 1922 Mackenzie King's Liberal government had introduced a bill into the House of Commons creating a Department of National Defence, with the new minister taking responsibility for the militia, and for military, naval, and air services. Civil aviation was included under the Canadian Air Force. The department came into existence on 1 January 1923, and the Air Board ceased to exist. Operations at air bases were not interrupted. Survey, photographic, and patrol work continued.

The Northland Beckons

Charles E. Taylor, manager of western development for Imperial Oil, was looking for ways to get personnel and supplies to the company's new well at Fort Norman, Northwest Territories. The 1,891-kilometre trip down the Mackenzie River from the end of the railroad at Peace River would take weeks by dogsled. Alternatively, the company could wait for spring breakup and make the trip by water. On the other hand, if the company chose to move equipment and supplies by air, it would take just days, and they could start right away. True, there were drawbacks: few fuel caches, untried landing conditions, unpredictable weather, unreliable planes, poor navigation systems. But there was one major advantage: time. Taylor decided to give it a try.

Imperial had purchased two Junkers, all-metal monoplanes with enclosed cabins. George Gorman and Wop May were hired to go to New York and fly them back. The progress of the two planes was followed closely in Edmonton. When one was damaged in Brandon during takeoff, the flyers decided that Gorman should stay with it while May flew on to Edmonton. There were other people with him, of course: a pilot and mechanic from New York, and the invaluable Pete Derbyshire. The monoplane landed in Edmonton at 5:20 PM Wednesday, 5 January 1921. A huge landing fire marked the May–Gorman aerodrome, and a welcoming party shook hands with May and his happy passengers. Imperial Oil was particularly pleased, and reaffirmed its plans to use the plane in the north.

Two weeks later, Gorman touched down with the second Junkers. He had been assisted by a strong tail wind that allowed him to maintain a speed of 210 kilometres an hour from Saskatoon. The most difficult part of the journey still lay ahead. For the flight north, Gorman would pilot one of the aircraft, and Elmer Fullerton, an instructor with the Canadian Air Force, the other. Imperial had already established a base at Peace River with a hangar and a small annex for the staff to live in.

The planes, named the Vic and the Rene, took off for Peace River toward the end of February. On board were seven men, including a dominion land surveyor and

Imperial Oil's hangar at Peace River, 1921. Courtesy Glenbow Archives, Calgary, NA-463-56

various geologists and surveyors from Imperial Oil, and 450 kilograms of equipment. They made Peace River in three hours flying time, and received a hearty welcome from the mayor and a group of citizens. At Peace River the engines were checked and a route was planned. The flyers decided to follow Hudson's Bay Company posts where the planes could replenish their oil and gasoline from stores the posts kept for their motor boats. They also decided to establish a fuel cache at the Hudson's Bay Company post at Upper Hay River, and each Junkers made a trip there carrying oil and 100 gallons of gasoline.

Finally, all was ready. On 24 March at 9:00 AM, Gorman and Fullerton departed for Fort Norman. They carried food for ten days: enough, they hoped, to weather an emergency. Within two hours they encountered heavy cloud and decreasing visibility, and were forced to alter their course, landing at Fort Vermilion instead of at the fuel cache. The blizzard lasted two days. On 27 March they left Fort Vermilion and made the trip to Hay River on Great Slave Lake.

The rest of the trip was difficult and dangerous. In separate incidents at Fort Simpson, the propellers on both planes suffered severe damage. The Rene was also damaged. The Vic was closest to being airworthy: all it needed was a propeller. With the resources of the Hudson's Bay Company and the nearby Roman Catholic mission, coupled with the practical knowledge of a woodworker and one of the flight's mechanics, a new propeller was fashioned from boards and a glue made from moose hide and hooves. It worked perfectly, but it was late April, and any attempt to reach Fort Norman was out of the question; there wouldn't be enough snow for their skis. They decided to fly back to Peace River.

Low on fuel, and with most of the snow gone from the ground at Peace River, the crew decided to land at nearby Little Bear Lake. They dropped a note to the

Imperial Oil's Junkers "Rene" and "Vic" before being flown north. Courtesy City of Edmonton Archives, EA-160-207

caretaker at the aerodrome in Peace River, asking him to send landing wheels and gasoline to the lake. With wheels on, they would be able to land at Peace River. Off to Little Bear Lake they raced. To their relief, the ice was still solid enough to allow them to land. When they checked their fuel, they found they had just half a gallon left—enough for only two or three minutes in the air.

In late May, with a new propeller and a set of pontoons, the Vic again took off for Fort Norman, and this time it made it. The plane returned to Fort Simpson, and in late August both the Vic and the newly repaired Rene took off for Peace River. On 24 August they landed in heavy drizzle. The cloud of misfortune that had beset the Rene was also present. On landing in the river, the plane hit a submerged obstruction, broke a pontoon, and overturned. As it drifted downstream, Gorman and the two others on board scrambled to safety. The plane finally came to rest on an island.

Despite the difficulties, Gorman and Fullerton had shown that the airplane could be a significant factor in the business of extracting resources from the north. Certainly, it was not without challenge. The harshness of the climate, the changeable landing conditions, and the distance between settlements and fuel caches dictated when trips could take place and what routes would be followed. The planes, while capable of long trips, had also shown their vulnerability. They needed regular supplies of aviation gasoline, as their engines didn't operate well on regular motor gasoline. They needed decent landing fields and long stretches of water for takeoffs and landings. Still, only the airplane offered "speed . . . and a clear highway in the air," and more and more people were beginning to think northern travel was now possible.

Have your picture taken with one of today's modern wonders! Courtesy Provincial Archives of Alberta, 72.462/2c

Out of the Shadows

Civil aviation slowed down in Alberta in the early 1920s. The barnstorming companies that had started up with such optimism after World War I either put their planes in storage or sold them. Exhibitions and passenger carrying couldn't pay the bills. When the supply of war surplus planes dried up, so did the dreams of small-town aviators who couldn't afford to buy those same planes new from the factory, or ride out the tough economic times.

As yet, there was neither a consistent nor a significant demand for air passenger or cargo transport in Alberta. The ground transportation network was working fairly well, and larger, more reliable cars made longer trips possible for business, vacation, or neighbourly visits. Lobby groups such as automobile clubs and the Good Roads Movement kept road building at the top of the provincial government's agenda. Railways were expanding in the 1920s as well. New lines into the Peace River country, from Edmonton to Barrhead, and from Edmonton to Fort McMurray made northern travel a bit easier. A well-developed system of river transportation served northern communities in the summer.

When the drone of an airplane was heard over Alberta, it usually had come from the High River Air Station, the busiest in the country. In 1922, the pilots had logged 267 flights totalling just over 1,075 hours. The station bought 4,000 gallons of aviation gasoline and 122 gallons of oil. Most flights were forestry patrols and used de Havilland DH4s. Another de Havilland was equipped with a camera for taking aerial photographs. Streams on the eastern slopes of the Rockies were recorded for the Department of Agriculture, and Banff townsite and other areas in the mountains and the Bow River basin also had their pictures taken. Back at the station, new workshops and storehouses were built.

In 1922, the aerodrome was launching pigeons as well as airplanes. As the *High River Times* reported, carrier pigeons were taken to various points along the railway and released, and they returned to the base every time. The plan was to send them up on forestry patrols, and if a pilot got into trouble he could release a pigeon carrying a distress message.

Wireless signals were also being sent during that busy summer of 1922. The Air Board spent over $4,500 on a "wireless telephone and telegraph transmitting and receiving cabinet" for the station at High River, another $1,000 for transmitting and receiving equipment and an amplifier, and a few thousand more went to a contractor to erect the "wireless buildings" and the concrete base and anchor blocks for the "wireless mast." In late June, members of the Royal Canadian Corps of Signals arrived to take over the running of the wireless station for the Air Board. The *Times* reported that testing of the "wireless telephony communication" between plane and base began immediately. The new system was designed to enable a pilot who spotted a forest fire to wire the location to the station or to a ranger's cabin so fire fighters could be dispatched immediately.

An article in the *Imperial Oil Review* explained how the device worked. On the plane the transmitter was powered by a small propeller placed on the wing. A copper wire with a weight attached to it was let out once the plane was airborne, and acted as the aerial. Signals could not be sent from the ground to the plane because the pilot could not fly with "ear pieces." The sound of the engine was the closest thing to an instrument panel these early planes had, and a pilot

Crowsnest Mountain taken during a High River station forest patrol. W. J. Oliver, Calgary Herald *photographer, accompanied the patrols on several flights.* Courtesy Glenbow Archives, Calgary, NA-2097-14

needed to be able to hear the engine at all times.

According to one flyer, however, the device was virtually useless. Pilots were too busy flying to operate it properly, and with cockpit temperatures hovering around -7° C, it was too cold to take off their gloves to run the key. Forest rangers were supposed to wait to receive any messages that might come through, but this interfered with their own patrols on horseback, and most of them didn't stay in their cabins.

If a fire was small, a pilot might be able to land and put it out himself. This was often the case when it had been set by a settler to burn trash, for example. These kinds of fires were frequently spotted in the foothills, and settlers must have been more than a little surprised to see a plane drop from the sky and put it out. No doubt the already embarrassed citizen was further chastened by a sharp rebuke from the pilot.

The air station carried out the same kinds of work in 1923, although the need for forestry patrols was reduced by lingering snow and wet weather. No patrols were undertaken until August. A survey of the Cooking Lake forest reserve was carried out from a temporary base at Wetaskiwin. For the first time, the station remained open during the winter, and cold weather flying tests were done. The pilots climbed to about 3,000 metres and simply flew around until they were convinced that flying in cold weather was not so dangerous. After landing, they recorded the air temperature during the flight, as well as any observations they had made about the aircraft. A final report, entitled "Operation of Aircraft and Aircraft Engines under Winter Conditions in Canada," concluded that winter flying was not difficult as long as "the pilot was properly clothed and the cockpit well screened."

Experiments carried out under the auspices of the Air Board by Professor Charles A. Robb of the University of Alberta added significantly to the knowledge of how engines and lubricants worked in cold weather. Robb also tested radiators, and determined the best alcohol and water mixtures for winter flying. Air-cooled engines were not yet common in airplane technology.

The de Havilland DH4s had held up well at High River, but by 1924 they were beginning to show the scars of four years of fighting wind, rain, and snow. Orders were placed for five new single-seater Avro biplanes. Smaller and lighter than the de Havillands, they also had a lower landing speed, making them ideal for patrol work. The aircraft were in High River before the end of the year, and ready for use by the 1925 season.

Patrol work in 1925 was hampered by forest fires in British Columbia that threw a blanket of smoke over the Alberta side of the Rockies. Flying was impossible on some days, and the smoke greatly reduced the amount of photographic work that could be carried out as well. Personnel at the station took up new ventures, including spore trapping around Morinville, Vegreville, and Wainwright for a study on wheat rust being carried out by the Department of Agriculture.

Meanwhile, southern Alberta flyers were doggedly trying to make aviation profitable. An announcement that Lethbridge and Waterton would be linked by air service in 1924 reunited H. H. Fitzsimmons and J. E. Palmer in another scheme. Southern Alberta Air Lines was based in Lethbridge and had one biplane, "Lethbridge II." The company adopted "Millions Now Walking Will Some Time Fly" as its slogan, and

Getting Started

Like early automobiles that had to be hand-cranked to get them started, early airplanes needed a manual spin of the propeller to get their engines going. Most flyers, as in the top photo, had to do it themselves. Sometimes a helping hand was required. The High River RCAF station had another solution. A Ford Model T was rigged up to transfer power from its drive shaft to the propeller. Called a "Huck starter," it made the work of the ground crew much easier.

Courtesy Glenbow Archives, Calgary, NA-2097-18, NA-2097-17, NA-2097-25

added, "Fly and the World Flies with You . . . Creep and You Walk Alone" to drive the point home. According to company advertising, the public, and businessmen in particular, were now convinced of the utility and safety of air travel. In July, Palmer flew the mayor of Lethbridge to Fort Macleod, and a Lethbridge businessman to his oil well near Coutts. As well as flights to Waterton ($20 one way, $35 round trip), the company offered advertising and photographic flights, and long- and short-distance charter trips. Special rates were available on flights to the oil fields, and to doctors answering emergency calls outside town. Unfortunately, "Lethbridge II" was damaged after a busy two days carrying passengers at the Pincher Creek fair in August, and the aircraft returned to Lethbridge rather ignominiously on the back of a truck. Fitzsimmons, discouraged, left the aviation business.

In the spring of 1924 Harry Adair, a wealthy rancher from Grande Prairie, bought a plane in San Diego. American pilots were flying it to the border, then Wop May would bring it through Edmonton to Grande Prairie. May described the plane's flight to an *Edmonton Journal* reporter as "sailing through the air as steadily as a big sedan rolling over a concrete pavement." Adair formed the Edmonton and Grande Prairie Aircraft Company with plans to barnstorm through the Grande Prairie area and expand from there. In late June, however, while taking off for points north, May was unable to clear Grande Prairie's telephone wire, sending the plane out of control. It crashed into an abattoir. May, Adair, and a passenger were unhurt, but the plane's wings and propeller were damaged, and new parts had to be ordered. By August the plane was back in the air, performing at Grande Prairie and Wembley, but there wasn't enough flying activity to sustain the business, and May returned to Edmonton with the plane. The company disbanded.

While that misfortune spelled the end of Adair's venture, it gave a gentle push to the development of aviation in central Alberta. The crash in Grande Prairie, Edmonton's mayor Kenneth Blatchford argued, pointed up the "forceful need" for good landing grounds. Wop May had been lobbying for an aviation field in Edmonton for some time, and city council was finally ready to take action. By the end of September 1924, the city engineer had inspected two sites with the commanding officer from the High River station. The site they settled on was the new Hagman Estate, a parcel of land between 118 and 123 avenues and 113A and 121 streets. It had been used by Jock MacNeill's air taxi service earlier in the decade, and Wop May was currently using it for his own flights to Grande Prairie. The commanding officer said he was satisfied with the field, declaring that its proximity to the centre of the city and to railway and utility services made it a good choice.

The land and the buildings on it, one of which had been used as a hangar, had come to the city in 1923 through non-payment of taxes. The city engineer's report estimated that $400 would need to be spent clearing brush, levelling the land, and seeding it before the site would be suitable for airplane landings. The city commissioners approved the area for airfield development, and in the spring of 1926 the $400 became available. J. B. Macdonald & Son were hired to do the work. By fall, three "landing courses"—one running north-south, one running east-west, and one running diagonally across the field—were ready. The city engineering department felt the hangar could accommodate three medium-sized planes.

On 16 June 1926 the field was licensed by the federal government as an official Public Air Harbour, the first such field in Canada to be owned and operated by a municipality. In November, Edmonton sent a letter to the Department of National Defence (DND) requesting its new airfield be called Blatchford Field in honour of the former mayor, now a Member of Parliament. The department agreed, and congratulated Edmonton on its farsightedness. The deputy minister expressed the hope that "the example by your city in establishing this flying field is one which I trust will be followed by every other city in the Dominion."

The air harbour was officially opened on 8 January 1927, a chilly Saturday. Two Siskin biplanes from the High River station landed just after noon. The pilots, R. Collis and C. H. "Punch" Dickins, and their mechanics were greeted by the mayor and other civic officials and presented with souvenir cigarette cases. The RCAF would be conducting cold-weather flying tests from the new air harbour, and the Siskins would be stored in the hangar when in Edmonton. The next day hundreds of people drove out to inspect the field, the planes, and the flyers.

The details of looking after the new air harbour began immediately. Someone had to be found to keep the landing courses level and seeded and free of brush in the summer. Not all of the land had been developed, and the city decided that the unused hectares could be leased. The initial lease was to run from 1 April 1927 to 31 March 1932, with the fees being $300 for the first and second years, $400 for the third year, and $450 for the final two. The lessee was responsible for any fencing; for cultivating at least thirty hectares; for harrowing, floating, and seeding (with seed provided by

The two Siskin biplanes from High River that helped open Edmonton's new Blatchford Field. One stayed in the city for cold-weather testing. Courtesy Glenbow Archives, Calgary, NC–6–11974e

the city) the landing courses; for keeping the grass cut on the landing strips (he could keep the hay) so they would be ready for planes at any time; and to ensure that no livestock wandered onto the landing courses or into the hangar. At last Blatchford Field was ready for business.

Everyone Into the Air . . . Again

Edmonton's optimism about its aviation future was based on its own air harbour. That optimism, the cumulative effect of developments in central and eastern Canada and around the world, was shared nation-wide. Throughout the 1920s, fledgling flying ventures had been struggling to get off the ground. They ranged from the practical, such as establishing airmail service in Newfoundland in 1921, to the spectacular, such as the attempt by Prest and Bach to fly from Mexico to Siberia (they landed in Lethbridge and Edmonton) later that year. Then in the spring of 1927, Charles Lindbergh climbed into the cockpit of his little Ryan monoplane on Roosevelt Field on Long Island and flew over to France. Others had crossed the Atlantic before him, but Lindbergh's was the first solo crossing. Indeed, he was the only aviator left standing from a group of flyers who were all trying to cross the Atlantic solo from the United States to France that spring, and claim a $25,000 prize. Lindbergh's flight returned some of the exotic to what was becoming an increasingly mundane activity. Concerned with the details of mail schedules and airfield width calculations, the aviation community was ready for a hero. Alone against the sky, the handsome aviator instantly became a brave and romantic figure in the imaginations of Americans. The rest of the world was also entranced, and the somewhat bewildered Lindbergh found himself embraced and lionized around the globe.

The Northern Syndicate's Vickers Viking at High River. The station served as a point of entry for customs. The staff also tested Alberta pilots, and inspected their planes for air worthiness. Courtesy Glenbow Archives, Calgary, NA-2097-19

Aerial exploration companies had been formed in central Canada by the mid-twenties to advance prospecting and mining projects in the Canadian Shield. By 1926, regular runs were being made into the Rouyn gold fields in Quebec, and the Red Lake district of Ontario. Alberta was another early participant in northern exploration. On 14 June 1926, the *Edmonton Journal* announced that "Prospectors in Large Vickers Biplane Will Cruise North Country." A Calgary organization, the Northern Syndicate, intended to survey prospective mineral and oil deposits by air. The group had already transported 10,000 gallons of gasoline to the Slave Lake area.

The survey trips, piloted by C. S. Caldwell, went successfully. Caldwell termed them a "roaring success" during an interview with a British aviation writer. The subsequent article published in a British aviation weekly, the *Edmonton Journal* reported, referred to Northern Syndicate's prospecting flights as "the outstanding event in commercial aviation for the year." According to Caldwell, the Vickers Viking flying boat easily carried five men, their tents, sleeping bags, food, and cooking equipment. Fort Fitzgerald was their supply source, but they cached "gas and grub" throughout the survey area. The plane was put into winter storage at High River. On its way there, it stopped on the North Saskatchewan River just behind the Macdonald Hotel in Edmonton, drawing many curious people to the water's edge for a look.

In September 1926, a seaplane piloted by J. Dalzell McKee, an American, with A. E. Godfrey of the RCAF acting as second pilot and navigator, crossed Canada from Montreal to Vancouver. McKee had needed to fly his plane to California, where the manufacturer was to make some modifications. By deciding to fly across Canada, he became the first person to do so using only one plane. McKee, grateful for the assistance he had received from the RCAF, the Ontario Provincial Air Service, and the civil aviation authorities, donated the "Trans-Canada Trophy" to the Department

Two RCAF Vickers Viking flying boats, the first planes to arrive in Fort Chipewyan, on Fraser Bay in 1927. University of Alberta Archives, 86-106-44; courtesy Brother Louison Veillette, OMI

of National Defence. It was to be awarded every year to a person who had made a significant contribution to the development of aviation in Canada.

The year ended on another strong note; James A. Richardson of Winnipeg incorporated Western Canada Airways Limited, becoming its president and only shareholder. By the end of that month, flights had begun into the Red Lake district in Ontario in a Fokker Universal.

Federal regulations governing civil aviation soon caught up with the activity in the skies. On 1 July 1927, civil and military aviation were separated into four branches, each remaining under the jurisdiction of the Department of National Defence. The RCAF retained control of military aviation, while civil operations now reported directly to the deputy minister of National Defence. The new Directorate of Civil Government Air Operations, which included bases such as the one at High River, the Controller of Civil Aviation, and an Aeronautical Engineering Division now looked after civil aviation.

The sixtieth anniversary of Confederation fell on 1 July 1927. Ottawa celebrated rather ironically by having an American, although it was none other than Charles Lindbergh, fly in for the festivities. Edmonton celebrated more modestly, with a Canadian: Lieutenant Paul Calder from the High River station entertained the crowds with loops, tail spins, and daring dives. He flew to Pigeon and Cooking lakes, and stunted over Edmonton golf courses and baseball fields. A pair of gauntlets fell from his spinning plane over the Mayfair Golf Club; it was later reported that they had been found, but not if they had been returned.

Just Like Old Times

Officials in the Department of National Defence wanted to increase the number of trained pilots in Canada. With few opportunities to fly, many World War I pilots were no longer actively involved in flying, and opportunities for newcomers to learn were expensive and not widely available. With commercial activity picking up, however, Canada would soon need more pilots. Moreover, given the part aviation had played in World War I, flying would undoubtedly be of critical importance in the next war. The Department of National Defence looked at the British system of flying clubs, and decided to establish that system in Canada.

The department offered to supply two aircraft free to any flying club that met certain requirements. These included having a minimum of thirty members prepared to study to become pilots, a decent landing field with a hangar, and a qualified instructor and air engineer. Instructors had to be approved by an inspector working for the DND. Each club also had to include some members who already had pilot's licenses. If a club purchased a plane, the department would match it. The DND also decided to subsidize pilot training. Clubs would receive $100 per student for the first thirty who earned their private pilot's license by taking lessons from the club. The clubs were not allowed to participate in commercial flying.

The federal government's decision to support aviation training was the most significant factor in the growth of flying in Alberta in the late twenties. The club plan put planes and money into the hands of local groups. It made low-cost training accessible to the general population, and it placed Alberta flyers once again close to the centre of aviation developments. The clubs, in turn, spurred the development or improvement of landing fields, and began to turn out skilled pilots and engineers.

Calgary and Edmonton organized clubs immediately. In Edmonton, a one-year membership cost $5, and the club carefully spelled out its goals: "to ensure the fullest development of Civil and Commercial aviation; to foster Education in Aeronautical Engineering; to encourage research, experiment and manufacture in the Science of Aeronautics; and to instruct as many pupils as possible in flying, enabling them to obtain Civil and Commercial pilot's certificates." Pupils completed the ground school course either at the Armoury or by correspondence. The course included instruction on the theory of flight, airplane and engine construction, rigging, and maintenance. The fee was $25. Once through ground school, a prospective pilot was required to take a medical examination before flying instruction could begin. Flying fees were set at $10 per hour of instruction, and $5 an hour for solo flying. "The total cost of tuition," club officials maintained, "will depend on the aptitude of the pupil," but they thought most students should be able to get their pilot's license after ten to fifteen hours of flying time.

Filled with veterans, the camaraderie and excitement of the clubs must have been reminiscent of the war years. World War I aces Fred McCall and Wop May became the first presidents of the Calgary Aero Club, and the Edmonton and Northern Alberta Aero Club. Instructors were appointed and ground schools began. By 1928, sixteen clubs were in operation across Canada, and another six were organized that year.

Soon the promised planes, de Havilland Cirrus Moths, began arriving. The government kept its word, and made sure the clubs fulfilled their obligations as well. In Calgary, for example, McCall left to take a position with a new commercial air service. His departure left the club without an instructor, making it ineligible for planes. Calgary's first plane arrived in pieces on the train that September, after the club's new instructor had passed a training course.

Early in 1928, the Edmonton and Northern Alberta Aero Club (ENAAC) reported that it had 100 pupils enrolled in ground school; they would graduate to flight training in the spring. The club became a strong promoter of aviation in Edmonton, urging the city to improve its airport, calling for the runways to be seeded and rolled, stressing the need for a larger hangar. The club asked that proper gasoline, oil, and maintenance facilities be provided, and landing lights be installed. The club also wanted a telephone and electric lights in the hangar. It was willing to take on the responsibility of running the airfield in exchange for a grant from the city to cover maintenance and some operating costs: $500, the club suggested, would be sufficient.

Edmonton and its aero club came to an agreement, with the club assuming control of all aviation activities. The city, in turn, assumed responsibility for maintenance and new construction at the airfield. The city agreed to install a telephone and electric lights in the hangar, and purchased the building from Imperial Oil for $250. The oil company, in its turn, handed the money over to the flying club. The Edmonton experience was not unique. Across Alberta, the flying clubs were closely linked with the development of airfields, another significant outcome of the flying club movement. In these early years, the Calgary and Edmonton clubs were consistently among the busiest in the country, and they tirelessly lobbied municipal officials for improved and expanded facilities.

It wasn't all work, though. In late June 1928, the ENAAC held an air show and dance. The club's new de Havilland Cirrus Moth was on display. Between dances, guests could stroll past World War I relics that included the machine gun purportedly used by Baron von Richthofen in his duel with Wop May, the wings from a Fokker shot down over France, and several bombs. Model planes made by club members were on display, too.

Calgary had also been casting about for a suitable site for an airfield, and sought Edmonton's advice on running a municipal aerodrome. Calgary asked for information on the charges made for landings and takeoffs by tourist and commercial aircraft, what services the city supplied, and the number of aircraft the field could accommodate. Airmen had been using a couple of fields around the city, but Calgary, encouraged by the Calgary Aero Club, wanted to stride into the air age with a federally approved airfield. On 28 September 1929, Calgary officially opened its licensed airfield with a spectacular air show put on in large part by the club. Twenty pilots and fourteen planes from as far away as Lethbridge and Moose Jaw entertained a paying crowd of about 6,000 and an equal number who watched from outside the field.

An altitude-guessing contest (the plane was 2,065 metres up when the siren went off), dead-stick landing competitions, stunt flying, and a relay race kept the crowd entertained. One competition involved the pilots taking rubber balloons into the air, releasing them, and then trying to shred them with their propellers. The event "necessitated some tricky handling of the aircraft," the *Calgary Herald* reported, but

Students and instructors in Calgary, 1929. Courtesy Glenbow Archives, Calgary, NA-3277-18

each pilot managed to score a hit. Then, as in years gone by, Fred McCall provided the most spine-tingling moment of the exhibition.

McCall was demonstrating an aerial bombing attack on the airfield, buzzing over the heads of the crowd, when he plunged to the ground and out of sight, as if he'd been shot down. The crowd rushed toward the spot, expecting to find splintered wood, twisted metal, and little of McCall. The ambulance that had been standing ready all afternoon raced across the field. But Freddy had planned it all. He hadn't crashed; he'd flown into a ravine at the eastern end of the field, keeping below the ridge. In a few seconds he soared skyward once more.

In the souvenir program produced for the occasion, the aero club reiterated its commitment to "co-operate with all who realize the part aviation is going to play in the development of our Dominion." The British Army and Navy store on 7th Avenue West advertised aviator's supplies, including commander goggles, Scully helmets, leather jackets, and British-made breeches. A local gas company proclaimed Calgary an ideal location for aviation because no smoke belched from chimneys. Natural gas keeps a city clean, the company boasted.

Red Deer wasn't far behind its sister cities in getting off the ground. In March 1930, the Red Deer Flying Club called a general meeting for anyone interested in flying, and in establishing an airfield. The Board of Trade endorsed the airfield idea and voted a $100 grant to help in its development. The flying club sold memberships at $10 each to raise the $1,000 necessary to operate the field for one year. By the end of April a lease had been signed. The club had about forty members and was hoping to recruit more. "Letters are received daily from air transportation companies and flying schools," the club reported, "which indicate the interest evidenced in the Red Deer airport." On 7 May, the *Red Deer Advocate* reported that the flying club would hold a "monster air circus" on the 24th of May to open the new airport. Members of the Saskatoon and Winnipeg aero clubs had been invited, as well as planes from companies operating in the west. Relay races, aerial dog fights, altitude and bombing contests, passenger carrying, and, of course, stunting were all on the program.

The Red Deer club by then had almost fifty members and was well organized.

Learning to Fly

Mrs Gladys Graves Walker received Private Pilot License Number 372 on 15 September 1927. Years later, at the age of seventy-seven, she recalled taking lessons with Wop May and Moss Burbidge at Blatchford Field. She would get up at 4:00 AM, take the street car as far as it went, and then walk the rest of the way to be at her lesson at 5:00. The landing courses were mainly gopher holes and bumps, she remembered. Insurance salesmen pestered her to buy accident and life insurance, but she bought neither. She recalls being the first woman Burbidge sent on a solo flight: "He was more nervous than I was!"

Committees had been struck to look after membership, finances, the airport, and entertainment. The airport committee was still working out such details as gasoline and oil concessions and airport fees, but the field itself was almost ready. Plans for the air circus were coming together. The club was selling stickers for $1 that admitted one car and four occupants into the grounds on the big day. Single admissions were 25¢. The three-and-a-half-hour program was scheduled to begin at 2:00 PM, with a dance to follow in the evening.

The afternoon was a tremendous success. At least 5,000 people, many from nearby communities such as Blackfalds, Sylvan Lake, and Penhold, crammed the grounds. A man from Kelowna won the $10 prize in the altitude-guessing contest. The winning time in the relay race—the planes started at the airfield, zoomed around the water tower and the provincial training school, and returned to the airport—was fifteen minutes and fifty-two seconds. The wind prevented a good bombing and balloon-bursting exhibition, carrying the balloons off before the pilots could catch them in their propellers. It had also prevented a number of planes from getting to Red Deer for the fair. For the flyers and the crowds, it must have seemed that the old barnstorming days were back again.

"An Absolute Necessity"

"Aviation is growing in Canada," the Edmonton Chamber of Commerce declared in January 1929, "and what may seem to be but a useless luxury today will become an absolute necessity in a very short time." The chamber was referring to the need for hard-surface runways, a decent hangar, up-to-date repair and service station facilities, markers, upgraded lighting and other utilities, and adequate parking for cars

Like many aspiring pilots in the late 1920s, Gertrude de la Vergne earned her license at an aero club. She took her lessons from the Calgary club, and wrote a column on aviation for the Calgary Herald *in 1929 and 1930. Behind her is Great Western Airways' Moth "Kit."* Courtesy Glenbow Archives, Calgary, NA-3277-7

and taxis at the Edmonton aerodrome. The chamber's declaration, though, could have been applied to virtually any aspect of aviation in Alberta. Just a couple of years before, no one could have predicted that a course on aeronautical studies would be offered at the Provincial Institute of Art and Technology in Calgary. Yet, in 1928, the course was offered. By 1930, evening courses were under way. The calendar explained that the forty-lesson course was designed "to assist those already engaged in industry by supplementing their practical work by instruction in the technical and theoretical branches of their trade." The cost was $10, and students were required to attend 90 percent of the classes and achieve a grade of 65 percent in order to get their certificates. The graduating class, according to the year book, could "fix any old crate on wings."

In 1931–32, a two-year day program was available for "men who are concerned with the proper repair and maintenance of aeroplanes and engines." This should lead, the calendar declared, "to interesting and remunerative employment." The institute also offered a three-month course in aeronautical engineering for private pilots who wanted to obtain commercial licenses.

In the first year of the two-year program, students attended lectures in rigging, aeronautical science, mathematics, drafting, physics, and English. They spent 225 hours in the automobile shop and another 225 in the rigging shop. The calendar outlining the course announced that the school had a Sopwith Camel airplane, two rotary and one V-type engine, and "numerous aerofoils" for the students to study. Aeronautical science covered everything to do with flying, from operating the controls to landing, taking off, and stunting. Students also studied the history and theory of flight, meteorology, and navigation. In the rigging shop, the students focused on removing and replacing the wings, understanding the bracing and control wires, and the repair and replacement of struts, ribs, and spars. They also learned about maintaining and repairing the wing fabric, fitting skis and floats to the undercarriage, and repairing skis, floats, and tires.

The second year included more detailed study of the engine and ignition system. The class dismantled and reassembled engines and studied the component parts and systems. Students learned to identify engine problems and how to install an engine and mount the propellers. Lectures included a discussion of airplane performance, air navigation, and how to tie down an airplane when it was windy. The ninety hours of lectures also covered aerodrome layout and management, regulations, lighting and beacons, log books, and aerodrome rules.

The burgeoning air industry needed trained people to repair and maintain the increasing number of planes now in the air. Flying clubs needed instructors and engineers if they wanted to qualify for federal assistance. The first graduates of Calgary's technical institute found immediate employment with the aero clubs or with one of the new companies offering flying services in Alberta.

The enthusiasm with which the federal government supported airfield development and pilot and engineer training was part of a plan to realize a trans-Canada airmail service. Parts of the system were already in place. By the fall of 1927, seaplanes were loading mail on and off ocean liners crossing the Atlantic, and regular winter service existed between a few isolated communities where other forms of transportation were difficult or unavailable. By 1928, ten official airmail routes were in operation. Unofficial routes, such as the regular runs made into Red Lake or Rouyn, also carried mail as part of their cargo.

In the fall of 1928, the push to delineate airmail and passenger routes in the western provinces began in earnest. "Big Aeroplane Developments Pending" the *Edmonton Journal* informed its readers in late August; mail would be winging its way across the west by next year. In early September, a Fairchild monoplane piloted by Squadron Leader A. E. Godfrey and Flight Sergeant M. Graham of the RCAF left Montreal on its way to Vancouver. The *Edmonton Journal* reported that the flight was not a stunt, but an inspection tour of air services and facilities, and a test for airmail routes. The flight path did not originally include Edmonton, but driving rain and growing darkness forced the plane down on the North Saskatchewan River in east Edmonton. Miners from nearby coal mines threw ropes out so the pilots could tie it down. Godfrey and Graham were taken to Kenneth Blatchford's home for dinner. The next day they flew to Lake Wabamun to refuel, and were off to Vancouver.

On the return flight to Ottawa the flyers landed at Peace River, narrowly missing telephone and telegraph cables that crossed the river. Hundreds of people reportedly came down for a look at the big Fairchild. On the next leg of the journey, smoke from forest fires reduced visibility to zero. The flyers had been following the Peace River and decided to try to get below the smoke to regain sight of it. As the paper reported, however, the "smoke made judgement of distance difficult . . . and in ten seconds from their first sight of water they struck the surface unexpectedly with such force

A Pitcairn Mailwing at the Calgary Aero Club's hangar at the landing field near the Banff Coach Road. The club used the shipping crates the plane was shipped in to build the hangar. Courtesy Glenbow Archives, Calgary, NA-3277-3

that the machine was crashed." The flyers and two passengers scrambled out of the cabin and swam to shore.

By luck, they had come down in sight of a trapper's cabin. They built a fire to attract the man's attention, and spent a week in his cabin before a boat took the party back to Peace River. A search plane had spotted the smashed plane. Little of it could be salvaged. The engine and the instruments were removed, but the plane, according to the *Peace River Times*, was "left in the river a hopeless wreck." The salvaged material, along with the flyers, was shipped back to Ottawa on the train.

That September, a little further south, Leigh Brintnell of Western Canada Airways Limited completed a survey flight over the proposed prairie airmail route. Flying a de Havilland DH61 Giant Moth, he covered the route from Winnipeg to Calgary. Strong headwinds forced him down at Leader, Saskatchewan, and Strathmore, Alberta, to take on additional fuel. On the 19th, he left Calgary for Edmonton and Saskatoon, where he spent the night before returning to Winnipeg on the 20th.

The federal government announced it would run an experimental daily airmail service in the west in December. On 10 December, Western Canada Airways got the run under way. One plane left Winnipeg heading for Regina and Calgary, while a second headed for Saskatoon and Edmonton. The return trips went from Edmonton to Saskatoon and on to Regina, and from Calgary directly to Regina. At Regina, the mail from both planes was combined and taken on to Winnipeg. The flights had been synchronized with the train schedules to facilitate mail pickup. The trip was successful, although not without problems. The Fokker Universal and Super-Universal aircraft had met the challenge. What had proven less satisfactory were the facilities along the route. It was now clear that good ground facilities, good communication between the ground and the aircraft, and accurate, up-to-date weather forecasts were essential if such flights were to be made regularly.

Ottawa decided the flights would be made, and the DND began making sure that landing fields across the west were ready. Over the next couple of years, the department acquired land, installed lighting systems, built radio and light beacons, and set up a weather centre in Winnipeg so that weather information would reach the pilots quickly.

Radio beacons were built at Lethbridge and Red Deer in August 1930. The *Red Deer Advocate* followed the developments. The call for tenders identified a radio beacon station and living quarters, each with a septic tank, and two storage buildings. The beacon was described in the paper as "a 5,000,000 candle-power revolving beacon" that would be used for night flying on the airmail route between Edmonton and Calgary. In October, the *Advocate* reported that Highway Lighthouse Company of Canada had won the contract to supply the beacons.

As well as major landing fields, the DND had designated a number of emergency fields to be spaced approximately fifty-five kilometres apart. In Alberta, plans were to locate them at Dauntless, Bow Island, Taber, Barons, Kirkcaldy, Gladys, Carstairs, Mayton, Red Deer, Ponoka, and Leduc. At both large and small fields in daylight, pilots were guided by orange cones spaced along the field. At night, smaller fields had floodlights or put out a line of flare pots. Most airports made do with boundary lights, red lights on obstructions, a rotating beacon, and an identification beacon.

An illuminated wind cone helped the pilot judge wind direction.

In October 1929, the mayor of Edmonton received an official letter from the Post Office: "I beg to inform you that the Postmaster General has authorized the inauguration of an airmail service across the Prairie Provinces between Winnipeg and Calgary via Regina, Moose Jaw and Medicine Hat, and between Regina and Edmonton via Saskatoon and North Battleford." Winnipeg–Calgary was clearly the main route, with the one to Edmonton designated as a feeder. The schedules of the Winnipeg–Calgary flights were designed to connect with mail trains. "In order to secure the greatest possible postal benefits from such a service," the letter continued, "it was found necessary to arrange for its operation during the hours of darkness." Lighting arrangements had not been completed on the Edmonton–Regina route, so that route would operate in daylight for the time being. Edmonton's assistance was asked in ensuring that the work of lighting the field was not delayed.

The Post Office had taken its time awarding contracts. Western Canada Airways was probably the favourite from the start, but Edmonton's Commercial Airways and Calgary's Great Western Airways lobbied hard for the Regina–Edmonton and Regina–Calgary routes. Commercial and WCA were also locked in a fight over northern mail territory, and each was determined to demonstrate its capabilities.

In January 1929, Punch Dickins left on an unofficial mail flight down the Mackenzie in WCA's Fokker. He was accompanied by T. J. Reilly, a postal inspector. Dickins took local residents up for joy rides between flights to Fort Simpson and Fort Resolution. WCA was also establishing a base at Fort McMurray. Commercial Airways of Edmonton responded by getting Post Office sanction for its weekly flights to Grande Prairie that had begun in May, and by flying to Winnipeg in six hours and forty-eight minutes later that month. "Breakfast in Edmonton, lunch in Winnipeg," the *Edmonton Journal* reported. The flight had proved beyond doubt that the Lockheed Vega was a capable little plane.

WCA began an advertising campaign in the *Edmonton Journal*. "Fly North," the

Come Up into the Sky with Me

Early flying companies used advertising to attract new customers. Some opted to stress how safe flying had become. An advertisement for Rutledge in September 1929 declared that the company stood for "safety, speed and service and the sane promotion of aviation. . . .The advent of the Rutledge Air Service has brought flying in Alberta up-to-date. The latest planes are used by skilled certified pilots." Others stressed both the excitement and the practicality of flight. A newspaper advertisement for Great Western Airways in July 1928 urged readers to "See Calgary From the Air. A fifteen minute trip in safety and comfort. Enclosed Stinson-Detroiter four-passenger airplane." The ad also noted that, "When roads are impassable we can get you there."

company urged, in a "Fokker cabin plane with spacious passenger capacity." Commercial immediately countered with a plug for its "usual flight" to Grande Prairie in a "comfortable Lockheed Vega cabin aeroplane." WCA's ad ran much more frequently, however, an indication of the company's greater financial resources. The two even competed in the mercy flight category. Dickins brought a man injured in a dynamite blast in Fort McMurray to Cooking Lake where the Howard and McBride ambulance took over, and Wop May rushed to Vegreville to bring a man with a broken back to hospital in Edmonton.

Commercial made its final attempt to win parts of the prairie airmail route in June when the firm merged with Great Western Airways of Calgary and a small company operating out of Regina. In announcing the merger, Vic Horner stated it would allow the group to make a unified bid for airmail contracts. The merger would not, however, affect the company's individual identities as far as commercial activities were concerned.

In late June, the Post Office announced its decision. The entire prairie airmail contract would go to Western Canada Airways. Horner and May reacted with anger and frustration, arguing that the decision would hinder the development of commercial aviation in the west. With guaranteed revenue from airmail contracts spread among several concerns, local companies would have been able to meet their basic operating expenses and use that base to look for additional business. The government's response was that it thought it was better if one company did the whole route, but then it argued that the amalgamation had hurt the smaller companies' chances. The Post Office, the *Edmonton Journal* pointed out, didn't explain this twist in logic.

The northern contracts were still up for grabs, the Post Office reminded the smaller companies. Commercial probably just sighed. It knew there was a market for mail, passenger, and cargo service in the north, but it also saw the long arm of an aggressive and well-financed company now well established in Alberta.

Millions Now Walking . . .

For Harry Fitzsimmons, 1924 had been a bit too early to see his dream come true. By the late twenties, though, the possibility to make a living through aviation seemed to be improving, and many people returned to the fledgling industry. The late twenties had seen a general increase in industrial capital and investment in Canada. After the doldrums of the early twenties, these modest economic gains meant more investment capital was available both from private investors and from the banking system. With the promise of airmail contracts and the steady money they would bring in, along with the potential for carrying cargo, hopeful operators could now raise the capital to buy planes and equipment. The experience of two Alberta companies, one in Calgary and one in Edmonton, illustrates the change.

In Edmonton in 1928, Wop May convinced a couple of his old flying buddies to join him in establishing Commercial Airways Limited. With $6,000, the company bought an Avro Avian, a small, open-cockpit, two-seater biplane. They wanted a small plane for stunting and joy riding at fairs in the summer, but it soon became clear that if they were to take advantage of the cargo opportunities opening up, they would need a more substantial machine. With this in mind, they purchased a Lockheed Vega from the factory in California and flew it back to Edmonton in February 1929. The Vega must have seemed like heaven. It could carry 275 kilograms of freight or seat four, it had a heated, enclosed cabin, and could go about 190 kilometres per hour.

The Vega's log book shows the flight path from Burbank through Las Vegas, Salt Lake City, Great Falls, and on to Calgary and Edmonton. For the first few months of 1929, the log book gives a good indication of the activities open to Alberta aviators during the period. There were paying passengers travelling to different locations, including Stettler, Vegreville, and Lacombe. There was passenger carrying of the old barnstorming variety, as indicated by entries for individual days when, for example, thirty-one flights were logged with a total of twenty-seven passengers. There were a few flights taking nurses and doctors to rural areas. In May the plane flew to Winnipeg for the opening of the city's flying field. In November, Archie McMullen flew the plane back to the Lockheed factory in Burbank for a new motor and propeller. After testing the new equipment, he returned to Edmonton in December.

Commercial Airways had acted quickly when it bought the Vega in early 1929, perhaps responding to Western Canada Airways' attempt to expand into Alberta. Commercial had been aware of WCA's intentions since early 1928. In May of that year two representatives of WCA based in Winnipeg had been in Edmonton looking at air facilities, and determining the feasibility of regular flights into the Peace River country and elsewhere in northern Alberta and beyond. Although the company did not find suitable facilities in Edmonton at that time, the representatives did indicate that they would soon be back to take a better look. Commercial had reason to be worried.

In Calgary in the summer of 1929, W. L. Rutledge was setting up Rutledge Air Service. As managing director, he hired a pilot, an assistant pilot, and an air engineer. Rutledge found the capital to get started in the local brokerage firm of Solloway

A horse-drawn mail sled contrasts with Commercial Airways' red fleet—two Bellancas and a Vega—in Fort McMurray for the start of the official mail run down the Mackenzie in December 1929. Courtesy City of Edmonton Archives, EA-10-2871

and Mills. Three new American Eagle A–129 biplanes were ordered from the factory in Kansas City, and on 24 July they arrived. The next day, the first of the three was assembled and taken up for a trial flight. By the end of July, Rutledge was offering two return flights a day between Calgary and Turner Valley. He was also involved in a charter service and in training pilots in Calgary.

In September, Solloway and Mills, who also had an Edmonton office, provided capital for Commercial Airways to expand. Rutledge and Commercial were now affiliated, but continued to operate under their separate names. With the new funding, Rutledge purchased a Curtiss Robin, and Commercial bought two Bellanca Pacemakers.

The impact of investment capital is apparent. No longer did companies have to depend on issuing shares and trying to sell them to friends and relatives as their predecessors had done just after World War I. Acquiring aircraft and hiring staff were now possible on a level adequate to get an operation up and running.

With the purchase of the Vega and the Bellancas, Commercial served notice that it was ready to bid for the lucrative trade ferrying cargo into and out of the north. With the purchase of the Curtiss Robin, Rutledge confirmed that his business would be based on pilot training and carrying businessmen and others on short hops. Calgary was too far south to tap into the northern trade, and with WCA controlling the long-distance prairie mail route, there wasn't much room for Rutledge. There was, however, a local market for the services he offered. With increased activity in the oil fields, especially from American companies, more businessmen needed to be taken on short hops to their wells. By the end of 1929, the company had logged 2,010 flights for a total of 852 hours.

In March 1929, Commercial Airways made the first flight in the Vega to Grande Prairie, landing on the field used by the forest patrol. A large crowd welcomed Wop May and his three passengers. The Board of Trade provided lunch. May tantalized Grande Prairie with talk of "first class service at economical rates" if he got the official mail contract, and if there were enough passengers who wanted to fly between Edmonton and Grande Prairie to support the route. May made several more trips to the area. Later that month he dropped off hospital supplies in Fairview before heading on to Grande Prairie, and on the return trip to Edmonton his passenger was Emilé Watson, the winter carnival queen. Miss Watson returned to Grande Prairie on the train, but was enthusiastic about her flight. She described it as "simply glorious," noting that "outside of two or three places, the plane glided along quite smoothly."

May continued flying to Grande Prairie throughout the spring, although he hinted that he was doubtful if regular flights would be able to continue. As it happened, Commercial Airways did receive the Mackenzie Valley mail contract, and on 1 November 1929 May moved his headquarters for the operation from Edmonton to Fort McMurray. The mail schedule involved weekly flights between Fort McMurray and Fort Resolution, a monthly flight between Fort McMurray and Fort Simpson, and a flight every two months between Fort McMurray and Aklavik. Commercial's first rate card, proudly announcing that it was an airmail contractor, gave the dates the northbound and southbound planes left Forts McMurray, Resolution, and Simpson, and Aklavik. The card noted that the planes stopped at "intermediate posts," and that passengers were welcome on all flights. A trip from Fort McMurray to Fort Chipewyan cost $35, from McMurray to Fort Simpson $175, and from McMurray to Aklavik $410.

Fulfilling the mail contract meant flying a gruelling schedule over some of Canada's most inhospitable territory. May immediately began establishing fuel and supply caches along the route. On 10 December 1929, the first flight down the Mackenzie began. It was difficult and dramatic. There was over three and a half tonnes of mail, and the planes had to take it north in batches. It took seventeen days in all. Blizzards and temperatures well below -40° C challenged the flyers at every stage. But they continued, dipping down to trading posts and settlements, delivering their cargo all the way to Aklavik.

Developments further south lacked the drama of the early northern trips, but they had all the hustle. In 1927, a Lethbridge man, C. B. Elliott, bought a plane and hired a veteran southern Alberta pilot, J. E. Palmer. The company, Lethbridge Commercial Airways, specialized in pleasure rides around Lethbridge and at nearby fairs. The next season Palmer left to join Purple Label Airline in Calgary, but not before

Loading a few of the mail bags into the Bellanca for the first run down the Mackenzie. Thousands of people took part in historic mail flights such as this one by sending letters on the flights. Special stamps adorned the envelope, and sometimes the pilot signed it too. Courtesy Glenbow Archives, Calgary, NA–463–49

he'd damaged the plane when he hit a rut taking off from Calgary. Elliott retired from flying for the remainder of 1928.

After using various fields around the city, Lethbridge finally settled on a piece of land in the north end. In 1927 the city applied to the federal government to have the field licensed. The perimeter was marked with white posts, and the centre of the field sported a large square outlined in white. It had been levelled, and Lethbridge citizens had been asked to stop using the trails that criss-crossed the land. The field was duly licensed as a "public customs air harbour for day flying."

In Calgary, brothers Frank and Fred Anderson decided to extend their auto business into flying, buying a Standard J-1 biplane. They were encouraged by one of their mechanics, Forin Johnson, who had earned his pilot's license and was anxious to use it. Johnson did the flying, mainly barnstorming and pilot training. In the spring of 1930 the brothers purchased a used Curtiss JN-4 and built three hangars at the airport. They also decided to become distributors for Eaglerock Airplanes, an American concern, and bought one for barnstorming. Unfortunately their enterprise had collapsed by the end of 1930.

Great Western Airways Limited, a product of one family's interest in beer and airplanes, got off the ground in Calgary in the summer of 1928. In 1901, Fritz Sick had begun the Lethbridge Brewing and Malting Company. When the Lethbridge Aircraft Company got going in 1920, Sick was one of the shareholders. Ads used by the brewing company often featured airplanes. In 1928, Fritz and his son Emil took over additional breweries in Alberta and Saskatoon, and Emil moved from Lethbridge to Calgary. He bought a Stinson Detroiter as a company plane, painted it bright purple, established the Purple Label Airline (named after a product of the Edmonton

Fuelling Great Western Airways' Stinson Detroiter. Courtesy Glenbow Archives, Calgary, NA-2097-62

brewery), and hired Jock Palmer and Fred McCall as pilots.

The plane could also be chartered by private groups. The charter service proved so popular that Sick established a new airline, Great Western Airways, that absorbed the Purple Label operation. Two more planes were purchased for pilot training. The company built a hangar in a field near the Banff Coach Road. It was a bumpy, rocky stretch of land bounded on one side by telephone wires—not an ideal landing location. But with their new Cirrus Moths, the company began training new pilots. The flying side of the business was kept equally busy: in 1928 the company recorded 3,000 flights, many of them to the mountains and to Turner Valley and the nearby oil fields. In 1929 Great Western bought a new de Havilland Cirrus Moth biplane and moved to the Calgary Municipal Airport, renting hangar space from the Anderson brothers. Business remained good, as again over 3,000 flights were logged and shareholders realized a healthy profit.

Aviation was big news. Newspapers reported the general goings-on at the airport and in the sky, and when students had passed their pilot's tests. When Joe Patton joined Great Western, the headline read "Calgary Pilot Joins Company." The article went on to report how he had "Shown early signs of being a first class pilot." Patton would be helping Jock Palmer with pilot instruction.

Pilot training formed a big part of Rutledge Air Service's business, too. From August 1929 to February 1930, the company ran a flying school in Medicine Hat. In

On 11 May 1929, two insurance men from Winnipeg on an aerial tour across the prairies misjudged the landing courses at Blatchford Field and crashed into a telephone pole and a barbed wire fence. The men were unhurt, but one of the plane's wings was smashed. Behind the broken wing are signs advertising Commercial Airways' pleasure flights. Courtesy Glenbow Archives, Calgary, ND-3-4675d

mid-January 1930, the company inaugurated ground school courses that were held at the high school in Red Deer. Rutledge continued operating its flying school and charter flying service in Calgary throughout 1930. He also supported the Calgary Model Airplane Club. Unfortunately, he lost his backers when Solloway and Mills were arrested for operational improprieties and barred from further trading in securities. Without financial backing, Rutledge amalgamated with Commercial Airways in late 1930. Title to the planes was transferred to Commercial, but Rutledge continued to run the Calgary operation, and still managed to log over 4,300 flights in 1930.

Commercial, riding high on the success of its airmail contract, had also been expanding during 1930. May's fleet of three brightly painted red planes, two Bellanca

With a spin of the propeller, May and Horner head north. The excitement generated by the trip is reflected on the face of Edmonton's mayor (far left), while the apprehension is apparent on the face of his companion. The Edmonton Journal *portrayed the trip in terms reminiscent of mediaeval Crusades: "Before and about them death, behind them prayers and hope. God speed them."* Courtesy City of Edmonton Archives, EA–81–26

Pacemakers and the Lockheed Vega, was spotted all over northern Alberta. The National Research Council and the Department of the Interior used Commercial's air service, along with prospectors, big game hunters, and the RCMP. The collapse of Solloway and Mills, however, coupled with the tenuous future of the airmail contracts as the Depression deepened and federal expenditures were cut back, made the future of Commercial look bleak as an independent small operator. Having absorbed Rutledge, whose financial base had been too small to withstand a major shock, Commercial found itself in the same position.

In May 1931, Commercial's chief competitor, Canadian Airways, assumed control of Commercial's aircraft and contracts. Some of the staff, including May, were kept on. The Rutledge planes were not included in the deal, and Rutledge once more found himself trying to make his own way in the aviation world. Revenue remained low. One plane was badly damaged in an accident near Stettler, and the remaining two were sold. The company logged only 225 hours in 1931, down from the 1,462 reported for 1930. The hangar was finally closed in December by sheriff's order, and Rutledge Air Service was no more.

The demise of Commercial and Rutledge was indicative of changes occurring in the aviation world in Canada. Larger companies were absorbing smaller ones all across the country. Companies needed a broad revenue base and sound financial backing as maintenance, equipment, and repair costs remained high, and the assured revenue of airmail contracts dried up. Although Commercial had initially been able to hold its own in Alberta, it did not have the resources to wait out a prolonged period of uncertainty. Western Canada Airways was rapidly becoming a giant in eastern and western Canada when Commercial succumbed to its offer. It was a difficult decision for Commercial; memories of the bitter competition between the two companies just a couple of years earlier were still vivid. But Commercial was only one of several companies WCA absorbed at this time, as it changed its name to Canadian Airways Limited and divided its interests into western and eastern operations.

The Rides of Their Lives

The message came into Edmonton from Peace River over the wire: diphtheria. One dead, six more gravely ill. An epidemic in the making. That had been the situation twelve days ago when the message began its trip by dog sled from Fort Vermilion. Wop May and Vic Horner readied their little open-cockpit Avro Avian as medical staff from the Department of Health prepared the anti-toxin serum to treat those who had the disease and the toxoid to administer as a preventative. The anti-toxin was wrapped in blankets and kept warm with charcoal heaters. If it froze, it would be useless. The flyers were wrapped in goggles, masks, and layers of wool and fur. If they froze . . . but no one wanted to think about that.

They left Edmonton about 1:00 PM on 2 January 1929, with charcoal heaters at their feet, chocolate bars in their pockets, and anti-mist paste on their goggles. It was bitterly cold on the first leg of the trip. Darkness and heavy frost on the wings forced an unscheduled landing at McLennan, where a landing field had been tramped into the snow by the townspeople and marked off by two lines of freshly cut spruce trees. The little plane came down safely, but material near one of the heaters was smouldering. Once that was dealt with, the two pilots were taken to the hotel to warm up. The next day they took to the air again, landing briefly in Peace River for refuelling. When they left, the temperature was below 0° C.

On the way to Fort Vermilion, a brick fell out of the heater at May's feet. Feeling the sudden increase in temperature, May had to reach down and replace the brick before the fuselage ignited. When they landed at Vermilion, chilled and tired, a musher was waiting to ferry the supplies to Little Red River, the centre of the outbreak. A tumultuous welcome awaited them: people cheering, sled dogs barking, and mushers leading their dogs in a frenzied show of welcome. Horner and May didn't return immediately, and this caused some anxiety among those who were waiting for news of their arrival. When their diminutive plane finally touched down again in Peace River, the *Edmonton Journal* was ecstatic: "With the rays of the setting sun gilding the wings of their machine in a final blaze of glory," they landed on the river. They spent a day going over the airplane, working in below-freezing temperatures and accepting congratulations and thanks from the community. On a runway pounded into the snow by members of the Alberta Provincial Police detachment, and with church bells ringing through the frosty air, they climbed into the sky. May waggled the Avian's wings in farewell, then started on the journey home. The flight back to Edmonton was as cold and dangerous as the trip up, but this time they were carrying emergency supplies in case they went down. Bacon, tea, bread, axes, and cooking utensils might keep them alive until help arrived.

The trip had caught Edmonton's imagination. It was a classic story of man against the elements, of human beings surmounting seemingly impossible odds to bring aid to a far flung settlement. It had all the ingredients for heroism: danger, long-distance flying, the possibility of death at every turn, and the aviator's altruistic

The Avro Avian pauses in Peace River on its mercy flight. Three steamboats, workhorses of summer transportation in the north, are beached in the background.
Courtesy Glenbow Archives, Calgary, NA–1258–62

disregard for his own safety. May and Horner had tossed their plane into the hands of fate in answer to a call of service. It was this ennobling veneer that was missing from those jobs—mail carrying, fire watching, prospecting—where flyers routinely met similar dangers. It was no surprise that May and Horner's war records featured prominently in accounts of the journey.

Throngs of cheering admirers met May and Horner at the aerodrome in Edmonton. The *Journal* estimated their numbers at 10,000, the *Calgary Herald* a more modest 5,000. There were more than enough, at any rate, to hoist the two heroes triumphantly on their shoulders in an outpouring of affection and pride. Sirens, gongs, and whistles filled the air, and one little girl was reported to have said that surely Lindbergh never had a welcome like this!

Congratulations and praise came from across the country and from around the block. Dr Helen MacMurchy, Chief of the Division of Child Welfare in the federal Department of Health, sent a letter of congratulations and asked for more details about the trip. Edmonton's First Presbyterian Church sent a letter of "deep appreciation": "We all admire your courage, and the devoted spirit which led you so promptly to follow the call of humane service. . . ." The city got a bit carried away in its plans to commemorate the flight, thinking it might purchase a plane. The cost, however, was prohibitive: $5000 (FOB Edmonton) for a Moth and $23,000 (FOB Edmonton) for a closed cabin plane seating four or five. And there was the tricky question of who would own it, the aero club or the city, or perhaps even the province. Perhaps a statue or a bronze miniature of the plane would be more appropriate. May and Horner were opposed to any such lavish commemoration, and in the end they were the recipients of a pair of gold watches and illuminated addresses of appreciation.

Why had they agreed to the trip in the first place? It seems clear that they wanted to do it, and they felt they could. With radio station CJCA blanketing the north with reports to be on the lookout for the flyers, and following what they called the trapper cabin route, perhaps they felt there was little chance they would not be found if the plane went down. They must also have seen it as an opportunity to demonstrate their willingness and ability to undertake northern assignments, for they were competing with Western Canada Airways for supremacy on some of the western routes at this time. There has been some suggestion that, had one of WCA's Fokkers been in Edmonton, it would have been asked to fly north first. But one wasn't there, and May and Horner grabbed their chance to show they could do just as well.

Several facets of this frigid trip, therefore, cemented their decision to purchase the closed cabin Lockheed Vega. The speed with which they left for California after their northern adventure points to a decision already having been made. Either way, they knew they hadn't made a mistake. For full-time northern flying, they needed a bigger, closed cabin plane.

Meanwhile in Calgary, Fred McCall was once again adding to his reputation as an aerial daredevil. J. C. Dallas of Calmont Oils Limited had decided to "shoot" one of the company's wells in Turner Valley. The oil wasn't flowing, and Dallas hoped a blast of nitroglycerine would make the difference. The closest nitro was in Shelby, Montana. Fred McCall agreed to fly it to Calgary in Great Western Airways' Stinson Detroiter.

International law prohibited the transport of arms, ammunition, or explosives by airplane, so McCall had to get special permission from Ottawa. Then he was off. On 22 February he brought back 100 quarts of liquid nitroglycerine, a box of dynamite sticks, and Charles B. Stalnaker, who would do the shooting. A small and apprehensive crowd began to gather at the landing field in Calgary around 5:00 PM, when McCall was due. The plane bounced when it landed, but there was no explosion. On his second trip, he narrowly missed ploughing into a snow bank while taking off from Shelby. Labelled "plucky" and "intrepid," McCall had done the job. In so doing, the *Calgary Herald* reported, he had proved how efficient and safe modern aviation was.

The 1920s had been a pivotal decade in the development of aviation in Alberta. Long distance flying reawakened both the romantic and the heroic aspects of flight. Mercy flights augmented this image. And so it was with renewed optimism that pilots and engineers turned their propellers into the sun. They didn't see the dark cloud slowly gaining on them from behind.

Sky Riders of the Plains

The Edmonton airfield was humming in the spring of 1929. Six planes were at the air harbour in early May, and the *Edmonton Journal* declared that the city had come into its own as a flying centre. By June, horses pulling ploughs, tractors pulling bigger ploughs, horses hauling harrows and seeders, and men swinging axes to clear brush were beginning to give the airfield what everyone hoped would be a surface as smooth as the felt on a billiard table.

Earlier that spring a crash had drawn attention to the state of the field. The flyers involved had remarked that, in a city the size of Edmonton, they had assumed a safe landing could be made on any part of the field. Two weeks before that, an American crew on a record flight from Kansas to Siberia and back had experienced a rough landing in Edmonton on their return flight. But it was May and Horner's dash to the north in January that had been most influential in persuading the city to upgrade its aviation facilities. The public, swept along on the romantic tide of the heroic possibilities of the airplane, was for the first time prepared to spend money on the airfield. That fall, city voters overwhelmingly passed a money bylaw approving major expenditures on the airfield and associated buildings, a project they had refused to support just a few months earlier. The promise of airmail didn't hurt, either. Edmonton needed a first-class field if excursions to the north were to be feasible, and if the city were to become a favoured landing spot for cross-country planes, mail planes, and the ever-increasing number of just plain planes that seemed to drop by on their travels.

In January 1929, civic representatives from Calgary, Edmonton, Lethbridge, Medicine Hat, Moose Jaw, Regina, and Saskatoon had met in Calgary for a day-long conference to discuss the development of aviation facilities in their cities. It was clear that landing fields would have to be increasingly sophisticated to handle the requirements of daily airmail flights, night flying, and the larger commercia⸍ planes now in service.

The mayors and administrators attending the conference wanted to exchange information and experience, and to develop some common policies in dealing with the problems of maintaining modern airfields, including servicing fields, night flying, landing and takeoff rates, and municipal liability in cases of accident. The conference ended positively, with an agreement that the cities should keep control of the airfields even if other organizations, such as commercial companies or flying clubs, ran them.

Edmonton's city engineer, A. W. Haddow, had been working on the plans for improving the airfield and its facilities. Those plans included a spacious, heated hangar, and room for an office and workshop. The plans to light the airfield met the standards set for primary landing fields on the prairie airmail route. A revolving beacon would indicate the field's general location, while boundary lights would outline the field and indicate how the approach was to be made, and floodlights would illuminate the field itself. Hazards would be individually illuminated, as would the wind cone. These improvements would bring the airfield up to Grade A standards as defined by the Department of Defence.

The department, realizing the extent of the improvements necessary to bring some airfields up to the mark, offered some assistance. It supplied one revolving beacon to each aerodrome, and agreed to pay for half the cost of purchasing the rest of the lighting. The cost of installation would be borne by those operating the aerodromes.

Early in February 1929, the city of Edmonton worked out the operational details with the aero club. The city would officially operate the field, with the aero club regulating air traffic. The city would build and maintain the field, the hangar, and any additional buildings required for administration, first aid, or meteorology. The city would also provide and maintain the lights. The aero club would pay for services to the hangar such as telephone, light, and water, and carry insurance on its contents. Private companies could erect their own hangars, but they would pay a ground rent to the city, and a use-of-field rent to the club. Oil companies could also build service facilities under permit on the field. All such private companies were responsible for maintaining their own facilities.

By the fall of 1929, tenders had been issued for the construction of the new hangar. It would be an impressive structure, anchored by a three-storey tower topped with the beacon. An inquiry office was tentatively planned for the tower's ground floor; the second floor was to be the airport office, and the third floor, closest to the sky, was the pilots' room. There would be room for meteorological equipment as well. The aircraft room would be twenty-five by thirty-one metres with a five-and-a-half-metre clearance, and the hangar doors would fold back into recesses when they were opened. Additional space was given over to more offices, a workshop, and a store room.

It was increasingly apparent that the airport could no longer be run on a volunteer basis by the aero club. Landing and storage fees as set by the Controller of Civil Aviation had to be collected; passenger tolls and various ground rental charges had to be paid; each incoming plane had to be noted in a registration book. With the increased activity at the Edmonton airfield, these duties took up more and more time. As well, each airfield was responsible for ensuring everything was in readiness when the airmail service got under way. Someone would have to be there in the middle of the night, every night, to turn on the lights.

In a report to the city commissioners in late November 1929, the city engineer pointed out how, in terms of traffic at the airfield, the activities of the aero club had been superseded by commercial operations. In 1928, the value of planes using the airport was approximately $12,000. By 1929, the figure had jumped to $250,000. Haddow had canvassed airports of similar size in the United States and found that the great majority were run by the cities in which they were located. He was now advocating that the city appoint someone as manager of the Edmonton airport.

It was only practical. The hangar, almost completed, was valued at nearly $22,000. The lighting was worth $9,300, and almost $6,500 had been spent improving the landing courses. Planes operated by Commercial Airways and Western Canada Airways used the field regularly, as did forestry planes from High River and a number of private planes. Airmail would only make things busier. A teletype machine was being installed to relay meteorological information along the mail route. Increased traffic in the air meant increased traffic on the ground in the form of cars and pedestrians. It would be the manager's job to regulate and oversee it all. The man Haddow proposed was J. A. Bell, vice president of the flying club, a Royal Air Force veteran, and a former city employee. Bell assumed the post in February, 1930. Hangar and office space would be rented to the aero club and to the commercial companies.

Edmonton's impressive new airport heralded a new age in air travel. Courtesy Provincial Archives of Alberta: Alfred Blyth Collection, Bl. 45

Prairie airmail would follow two routes, although it shared one plane on the Winnipeg-Regina-Winnipeg segments. The southern route went from Regina through Moose Jaw and Medicine Hat up to Calgary. The northern route branched off at Regina to North Battleford and on to Edmonton. Construction at airports and emergency landing fields across the west was proceeding at a frantic rate as the launch of prairie airmail service got nearer and nearer. A shortage of beacons and lighting equipment delayed matters somewhat; while lighting on the southern route would be completed by the March 1930 starting date, facilities along the northern route would not. Those flights would have to be made during the day until the lighting had been installed at all points along the route.

Test flights carried out in February were hampered by bad weather. Fog, and low cloud cover that in one case forced a plane to within twenty metres of the ground, interrupted the flights. Inconsistent weather across the prairies was also a problem. Snow cover at one city might dictate that the plane land on skis, while bare ground at the next would dictate that the plane land on wheels.

Edmonton, however, was ready. The upgraded airport was completed in February, and a trial run with the lighting system found everything in working order. On the Saturday before the inaugural flight, the Fokker rose into the night air above Edmonton, blinking its landing lights. The red and green navigation lights were also clearly visible, along with blue flame from the exhaust pipes. It was an impressive sight for the hundreds who turned out to watch.

Western Canada Airways and the Post Office had been publicizing the inauguration of airmail service. Businesses had been encouraged to decorate their windows

Bags of mail for the first prairie airmail flight are loaded into a truck at the Edmonton Post Office for the trip to the airport. Courtesy Provincial Archives of Alberta: Alfred Blyth Collection, Bl. 39/2

with air-minded displays, and, as the *Edmonton Journal* reported, "propaganda through the mails has been distributed." The paper tried to remain blasé about the whole thing. The city had already witnessed the airmail flight down the Mackenzie. "The new service," the *Journal* argued, "presents itself as less of an innovation to Edmontonians than to the citizens of other points who are brought into contact with it."

The Post Office made sure the public was well-informed on how to take part in the event. It was quite simple: address an envelope to yourself and put a 5¢ stamp on it (it wasn't necessary to use an airmail stamp). The route was to be indicated on the upper left corner: for example, "Via Edmonton–Winnipeg flight." Put these envelopes in a larger envelope addressed to the postmaster in Edmonton, with "Airmail enclosed for first flights" appearing prominently on the outside. The postmaster would see to the rest.

The participatory nature of these first airmail flights illustrates how accessible early aviation really was. Any citizen could drive out to the airport and walk around the Siskins that had just landed from High River and chat with the pilots, or rush onto the landing field when May and Horner returned from the north. Planes that landed on the river could be inspected at close range by simply scrambling down the river bank. And at fairs and stampedes, anyone and everyone could crowd around the planes and be taken for jaunts across the country. Now people could send as many letters as they wanted on this first flight, and those letters would come back to them, stamped and officially noted as having been on the flight. The accessibility of aviation explains much of the fascination with flight that resurfaced during the later twenties. The public wasn't just watching, it was participating.

As was to be expected, however, some of the more traditional-minded residents of Alberta were having a hard time adjusting to all the activity in the air. Many newspapers carried the report of an incident in the Nobleford area in November 1929: "Cow in Pasture Dies of Fright as Plane Passes." She had been seen to "dash madly across the field, and suddenly fall to the earth" as the airplane passed overhead. A ruptured blood vessel brought on by fright was blamed for the animal's demise.

In Search of Fireflies
Western Canada Airways took to the skies on the airmail routes on 3 March 1930. In Edmonton, the plane took off with more than 450 kilograms of mail and express, watched by an enthusiastic crowd. The *Edmonton Journal* had placed a microphone on the roof of its building. The roar of the Fokker's motor was picked up and transmitted over CJCA radio as the plane flew low over the city. A short time after the plane had left, a telegram arrived from Fort McMurray; the reception of the roaring motor, Wop May and friends reported, had been perfect.

Flights in and out of Edmonton had taken place in daylight. On the southern route, the mail travelled at night. A *Journal* correspondent rode the night flight from Moose Jaw to Calgary. "The progress of the ship," he reported, "becomes that of a midnight bird in search of fireflies." Droning steadily, the Fokker swept over the blinking beacons of the emergency landing fields as each appeared, a dot of comfort and familiarity in the dark landscape. As the plane approached a landing point, the pilot flashed the landing lights, described as "powerful beams from under the sides of the wings." This signalled the airport operator to turn the floodlights on the landing field so the pilot could see to land.

Approaching Medicine Hat at 4:00 AM, everything went according to plan. First the beacon came into view, then the lights of the city appeared, then the aerodrome boundary lights. The pilot flashed his lights, the floods came on, and the plane swooped in for the landing. Unfortunately, fog, snow, and cloud delayed the takeoff until much later that morning.

· The eastern leg of the southern route got off to a good start, too. Pilot Herbert Hollick–Kenyon landed and headed to the "rest station" for coffee and sandwiches, snugly bundled up in what the *Lethbridge Herald* described as a "fur overall suit." The plane was refuelled with fifty-six gallons of aviation gasoline, and was quickly back in the air, winging its way to Moose Jaw.

Medicine Hat had prepared well for its role in the airmail drama. The field had been carefully levelled and rolled. The eighteen-metre steel beacon tower was up and ready, and all the lights were in place, including red lights on the three tallest buildings in the nearby exhibition grounds to warn pilots of the danger. The airfield was well equipped with service stations: Imperial Oil had erected a building and pump, as had North Star, which had the contract to supply the airmail planes. Newspapers in southern Alberta described it as "one of the finest

An interested crowd watches the mail being loaded into the plane for the first prairie airmail flight out of Edmonton. Courtesy Glenbow Archives, Calgary, NA–1258–79

landing fields in Canada, and perhaps the best in Western Canada."

Medicine Hat residents had been encouraged to participate in the first flight. "Speed Your Mail By Air," they were urged, and a local store adorned its windows with a collection of first airmail flight covers from around the world, a model plane, and an action scene of a plane flying through clouds. The publicity worked. Several thousand flight covers were waiting at the Medicine Hat Post Office for the opening day of the prairie airmail service.

At the end of the year, the Post Office claimed the prairie airmail flights were operating at 98 percent efficiency, flying over 4,000 kilometres every twenty-four hours. The time saved in delivering a piece of transcontinental mail was twenty-four hours.

Once the Winnipeg-to-Edmonton and -Calgary flights were running, the Post Office looked to expand the service to Vancouver. Potential routes across the Rockies had already been surveyed, following the CNR rail lines from Edmonton, the CPR from Calgary, and through the Crowsnest Pass from Lethbridge. Rutledge Air Service out of Calgary flew the survey flight over the Rockies in its Curtiss Robin, hoping it would help them secure the contract to carry the mail on that route. These flights were particularly hazardous because the trails were inaccessible during the winter, and if the flyers went down, there would be no way to get to them. Still, the flyers, W. L. Rutledge and Percy Payne, took little food with them; they must have been confident of success. They did make it to Vancouver, but fog forced them to retrace their steps and land in Merritt.

Flyers from High River surveyed the southern route through the Crowsnest Pass. This route was finally selected as the safest because it had the most potential for emergency landings. It was also the shortest. In April 1930 the Post Office called for tenders from "flying concerns" to run a three-month trial on the Crowsnest route. The planes had to be able to travel 193 kilometres per hour in order to handle the buffeting winds they would encounter. Survey routes were also carried out on the potential Lethbridge–Calgary–Edmonton route, and on the Lethbridge–Medicine Hat route to facilitate the construction of emergency fields.

Lethbridge was selected as the junction of the north-south and east-west expansion, spurring the city into the airport improvement business. Early in June, Lethbridge burgesses voted on a money bylaw to spend $20,000 for lighting, a hangar, and other improvements to the airfield. The bylaw was passed by a "huge majority" according to the *Lethbridge Herald*.

Lights were ordered, and a contour map of the field drawn up. Tenders were issued for a hangar large enough for the mail plane and several smaller planes. Living quarters for the airport manager would also be constructed. The field would be fenced with wire, and the usual boundary lights and floodlights installed. By early August, the frame of the hangar was up, the cable for the boundary lights was being laid around the field, and the fencing was in place. Hollows and rough spots identified in the contour survey were filled in with dirt from a construction site nearby. The federal inspector was pleased with the progress of the improvements.

A wooden structure went up to house the floodlight, but it was decided to mount the revolving beacon on the old water tank just north of the airport, and the whole structure was painted with yellow and black stripes to make it more visible to pilots. Municipal grading equipment filled in hollows, and erased the remnants of old trails across the landing field. By the end of November transformers had arrived for the beacon, and trenches were dug for power, telegraph, and telephone wires. The latter were laid underground to keep the area within 300 metres of the beacon free from obstructions.

Emergency fields had to be developed along the new routes. The air route to Vancouver would enter the Crowsnest Pass from the Alberta side at Turtle Mountain, and guiding lights were placed at this entrance. Coleman had been chosen as an emergency landing site on the route, but the old forestry landing field was not large enough to accommodate the bigger mail planes. A new site was selected west of the old one, parallel to the highway and rail line.

On the rest of the southern prairie routes, emergency fields were established at Bow Island, Taber, Vulcan, Kirkcaldy, Barons, Dauntless, and Gladys. In Vulcan, 500 posts were ordered from the Co-op for fencing. A field six kilometres southwest of the town would be rolled and seeded, and electric lighting installed. At Taber, members of the Board of Trade laboured in the cool fall air removing stones and other hazards from the landing field before the runways were graded, levelled, and seeded. At Barons, work was delayed by an epidemic that quarantined the town.

Lethbridge couldn't wait to try its new facilities, and in early October Charles Elliott made the first night landing at the airport. With the boundary lights and floodlights on, he took off and landed, the lights on his own diminutive craft twinkling like stars. The next night, 3 October, a night flying exhibition was staged, and many Lethbridge residents drove out to watch the planes. Some went up to see their city for the first time from the air, and tried to guess what the different lights were. One set in particular had them stumped, as the *Herald* related: "Pilots and passengers were puzzled for some time as to what a row of lights towards the central part of the city represented, it finally being decided that the 'Wee MacGregor' miniature golf course was the cause of the illumination."

The date set for Lethbridge's inauguration into the prairie airmail circuit was 16 January 1931. By 12 January 10,000 letters were waiting at the Post Office. To the consternation of local officials, few were from the Lethbridge area. Philatelists and aviation enthusiasts from as far away as Cuba, Britain, the Philippines, the southern United States, and every province in Canada had sent letters. The excitement grew, though, and each new development was closely followed. The contract to haul the mail from the Post Office to the plane was signed. A carload of "ethylized aviation spirits" arrived for North Star. The *Herald* reported that it sold for 8¢ more per gallon than regular gasoline, "indicating that it must be far superior to the usual fuel." Then, to the relief of the Post Office, Lethbridge residents came through. By 14 January, Post Office employees were doggedly stamping their way through a flood of local letters. There was no escape even at the movies: 10¢ got you in to watch a special Saturday morning showing of William Boyd in *The Flying Fool*, a "smashing drama of the skies."

Remembering Those First Airmail Flights

An easy way to become part of the excitement of early aviation was to send a letter by airmail. The letters often carried commemorative stamps called cachets, and were sometimes signed by the pilot or a Post Office official. A flying company that inaugurated an airmail route sometimes had its own stamp printed, and sold it to help defray the cost of the flight. These stamps have been reproduced by permission of Canada Post Corporation.

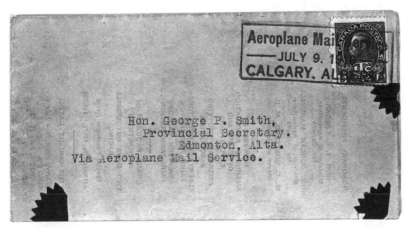

One of the 259 letters Katherine Stinson carried on her flight from Calgary to Edmonton in 1918. It was sent by E. L. Richardson, the manager of the Calgary Industrial Exhibition. "As this is the first time in Western Canada, and the second time in Canada that mail has been carried by aeroplane," he wrote, "I thought you might like to receive a letter delivered in this way."
Courtesy Keith R. Spencer, Edmonton

Cachets, often made locally by stamp enthusiasts, celebrated a variety of aerial events.
Courtesy Stanley G. Reynolds, Wetaskiwin, Alberta

This envelope was carried on the first Edmonton-to-Athabasca mail flight, in 1931. It has a locally made cachet on the front and a stamp printed by the carrying company on the back. Companies delivered to areas not accessible by land or ship under contract with the Post Office. Courtesy Stanley G. Reynolds, Wetaskiwin, Alberta

Airmail envelopes and stamps were available almost from the beginning. Airmail stamps, however, were not strictly required for these early flights; ordinary stamps would do. Courtesy Keith R. Spencer, Edmonton

Cachets commemorated final flights, too. This envelope was sent on the last prairie airmail flight from Edmonton to Winnipeg. Courtesy Stanley G. Reynolds, Wetaskiwin, Alberta

A cachet issued by "Bob of the Northland," a stamp dealer and promoter in Edmonton. He also issued a set of twelve stamps depicting Alberta's industries. Courtesy Keith R. Spencer, Edmonton

A cachet developed for one of the first trans-Canada mail flights. Courtesy Stanley G. Reynolds, Wetaskiwin, Alberta

Everything was ready. A pilot toured the fields and declared them satisfactory. A last-minute change in schedule had dictated that the first plane into the city would be the westbound plane from Winnipeg arriving at 2:45 Thursday morning, 15 January, not the eastbound plane from Calgary scheduled to land at 10:25 that evening. The city decided to continue with its original plans and treat the eastbound plane as the inaugural flight. Then, for perhaps the only time in the airmail schedule, bad weather conspired to set things right. A blizzard kept the westbound plane in Saskatchewan, making the eastbound flight later that day the true first flight.

Almost 4,000 people waited at the airport, their cars lining the field. They had been warned in the paper that no one would be allowed onto the field, "the possibility of accidents happening causing this definite ruling to be made." Sporadic cheers punctuated the cold night air, then great cheers went up as the plane's lights were spotted, and continued as pilot Herbert Hollick–Kenyon guided the Fokker down onto the field at 10:15 PM. He taxied to the hangar, where the mail bags were transferred. There was a brief pause for official greetings from civic, Post Office, and federal officials, some photographs were snapped in the glare of the floodlights, and then Hollick–Kenyon took to the air again. He returned almost immediately for a quick engine repair, but then he was back in the sky and on to Medicine Hat.

Hollick–Kenyon later reported that he had been able to see the Lethbridge beacon from 130 kilometres away. The flashing light, bouncing off the clouds, had puzzled some residents of southern Alberta, who wondered if perhaps it was lightning.

A *Lethbridge Herald* staff member accompanied Hollick–Kenyon on one of the early Calgary-to-Lethbridge flights. He noted that the pilot wore "fur-lined garments" and sat in the open cockpit, while the reporter reclined on leather-upholstered seats in the toasty warmth of the heated cabin. The glare of the gas flares from Turner Valley was perhaps the most spectacular sight as he enjoyed what he called "'Fly-By-Night' Thrills High Over Vales and Hills."

The world seemed to open up for the people of Lethbridge. When a Winnipeg-to-St Paul airmail route opened on 3 February, Lethbridge could claim that it was now linked by air to such exotic destinations as Buenos Aires and San Francisco. Soon, it was hoped, all of Canada would be joined in one big airmail route, stretching from east to west, and north to south. From sea to sea to sea, friends and family drew closer and news became a little less stale as planes and determined pilots cut hours and days from mail schedules.

Weather remained the biggest problem. As Commercial Airways noted in its winter 1930 schedule to points north, "The schedule is tentative, and rigid adherence to it will depend upon weather." Fog, high winds, snow, temperature changes that caused the wings to ice: these were some of the conditions that put planes off schedule, or worse. A pilot and his two passengers on a mail flight in September 1930 crashed to their deaths in a foggy field near Southesk, Alberta, while just west of Winnipeg two people died when a mail plane went down in February 1931. The federal government embarked on an ambitious program building meteorological stations along the airmail route, so weather information could be transmitted quickly and accurately from one point to the next.

The mail had just taken to the air, it seemed, when it was brought down to earth again. Canada in 1930 was still optimistic that its economic woes might be short-lived, but by 1931 it was clear the turn-around would not be soon. Airmail routes were cancelled, expansion was abandoned. The route to Vancouver was put on hold. Thousands of dollars were slashed from the Post Office's planned airmail expenditures. On 30 March 1932, prairie airmail service was cancelled altogether.

The Changing of the Guard

The training was rigorous, and it began when they were just twelve weeks old: pigeons were still being trained at High River in 1930. Not every one would be accepted into the air force. It took three years of study and endless hours in the sky before they were ready to take on their full responsibilities. Wearing metal tags, which bore their individual number and identified each bird as belonging to the High River RCAF station, they were shipped in wicker baskets to station masters along the rail line. When a station master received a basket, he released the birds. They started at two or three kilometres, and by the end of three years they were undertaking flights of 900 kilometres or more, returning to High River every time. The birds could achieve 120 kilometres an hour with a tail wind, 60 with a head wind. Notes could be placed in metal capsules attached to their legs.

The pigeons were just one part of the forestry service headquartered at High River. The single wireless mast had been dismantled in 1928 and replaced with two lattice steel masts as an improved system of wireless communication was put into

The forest fire patrol station at Grande Prairie, 1928. The wings of Moth airplanes folded back, making it possible for them to fit into smaller hangars. Courtesy Glenbow Archives, Calgary, NA-2097-37

operation. The masts and the radio station were moved to one side of the airfield. In another effort to help aviators throughout the west, these stations broadcast local weather reports and passed on weather information from stations in other provinces.

The RCAF established a temporary base in Grande Prairie in 1928 to carry out forest patrols in the Peace River district. Sub-bases for landing and refuelling were built at Pincher Creek and Rocky Mountain House. Most planes now carried radio transmitters, and the Department of Defence reported that "reliable voice communication" could be maintained from distances up to 320 kilometres.

Forest patrols continued from High River and Grande Prairie in 1929 and 1930. A civilian accompanying one of the High River patrols reported his experiences to the *High River Times*, describing the skill of the pilots, their dedication to the job, and the incredible scenery. The aviators became part of the local communities, and the newspapers often reported their social activities. The spring of 1930 found the High River flyers practising softball, the *Times* reported, "determined to lead the league again." They were also busy with their usual duties of airworthiness inspections, airfield examinations, and testing the Moths before they were distributed to the flying clubs.

Changes were afoot, though, for these outposts of activity. In late 1930, the

An air force pilot, heading off on a fire patrol, climbs into a de Havilland Moth at the airfield at Grande Prairie. Courtesy Glenbow Archives, Calgary, NA-2097-36

western provinces assumed full control over natural resources in their jurisdictions. This meant that the federal government would no longer look after the forest reserves. On 1 April 1931 the station at High River was put on "care and maintenance" status under the supervision of the Winnipeg station, and no flying operations were carried out. Grande Prairie became simply a storage location for aircraft. The new provincial forestry department decided that the view from lookout towers was good enough, and did not anticipate using airplanes. In early 1931 the High River aviators began receiving assignments to other stations. Most went to Borden, Ottawa, or Winnipeg. They were sad days for the little town, both for the friends now leaving and for business lost. As the airmen left to start their new lives, knots of people gathered to wish them well and see them off.

Sky Riders of the Plains

Ed Castor scanned the sky and decided, "Why not?" He was eighty-two, and his feet had been firmly planted on the ground for all of that time. It would probably be his last chance for a flight, he thought, until "they give me wings and tell me to go it alone." He declined a helmet and climbed into Charles Elliott's Waco in a dusty field near Claresholm, in late October 1930. He was happy with his decision as he watched the roads shrink into white strips from the air.

Many Claresholm residents took to the air that Wednesday. Five planes from various companies were on a tour of southern towns, performing stunts and giving aerial tours of the countryside. Over 1,000 people had come out to watch the stunts at the Claresholm "Air Circus." It was an appropriate name for the show, as the little planes swirled and dipped like trapeze artists under the big top of the endless western sky. A well-attended dance closed the festivities in Claresholm, dubbed a "thoroughly air-minded town."

Aircraft on the first Alberta Air Tour at Pincher Creek, 1930. The hangar and airfield had previously been used by forestry planes. Courtesy Glenbow Archives, Calgary, NA-3277-72

The planes had met a couple of days before in High River. Charles Elliott from Southern Alberta Air Lines, and seven planes from Rutledge, Great Western, and the aero club in Calgary then started winging their way south, doing shows along the way. From High River they travelled to Nanton, Claresholm, Pincher Creek, Macleod, and Lethbridge. In Lethbridge they advertised a novel way to calculate fares: a scale was hauled onto the field, and potential passengers paid 1¢ for every pound (about half a kilogram) they tipped the needle. An air queen was selected at each stop, and the following February the queens were treated to afternoon tea and a dance in Calgary, and a tour of the city from the air.

Air circuses were popular in the first couple of years of the 1930s. Planes from various companies put on shows together, often using the penny-a-pound rule for passenger rides when the stunting was over. One air circus in Cardston in the spring of 1931 featured a parachute jump, formation flying, airplane racing, and stunting. The Cardston Silver Brass Band was on hand to rouse any lagging spirits. Twenty-five cents got you into the grounds, and the cent-a-pound deal got you into the sky.

Often the local board of trade was a sponsor. Historically they had been in the forefront of town boosterism, enthusiastically supporting opportunities for growth while shaming anyone they felt to be standing in the way of progress. Air circuses were one way of boosting their town by demonstrating how air-minded, and hence modern, the citizens were. A good turnout from the surrounding area also presented business opportunities, as visitors needed cooling drinks, nourishing snacks, perhaps a meal in a café, and gasoline for the car ride home. Being air-minded in 1930 meant that your town was on the map.

The Cardston air circus was part of the All-Canadian Air Tour organized in the spring of 1931 out of Calgary. Well-known aviation personalities such as Freddie McCall, Joe Patton, and Ernie Boffa spent much of the summer touring southern Alberta and other western provinces. In Red Deer, for example, the aero club spent a good deal of time preparing for the show. Committees were struck to look after tickets, parking and grounds, and advertising while members of the Women's Institute took care of the refreshment concession. Free rides would be given to the person who sold the most tickets, as well as to the holders of tickets drawn at the show. "For Real Thrills . . . the Air Circus will be Hard to Beat" the advertising promised. On the big day, 1,000 adults and 500 children watched nine planes loop and dive overhead. Parachute jumping kept the crowd on edge, especially one jump when the parachute didn't open. Cries of alarm turned to laughter as people realized that the jump had been made by a dummy. Many people went up for rides, and the afternoon was a grand success.

The tour stopped at small towns across Alberta: Taber, Cardston, Granum, and Milk River were all air capitals for a day. Lethbridge and Calgary were also treated to the All-Canadian Air Tour, in Calgary's case as part of the Stampede.

The first large air tours came to the prairies in 1930 and 1931. The National Air Tour for the Ford Reliability Trophy had been crossing the northern United States since the mid-1920s. Designed to showcase the newest airplanes and educate the public about the reliability of air travel, the tour hopped from city to city on an advertised schedule. The planes were carefully timed between cities, but the fastest didn't necessarily win the race. A complicated set of calculations also took into account the plane's horsepower, load, and its ability to take off and land quickly.

In 1930, Edmonton, Calgary, and Lethbridge were included for the first time. In Edmonton, excitement mounted for the tour's arrival on 17 September. The pilots and their planes were featured in newspaper articles. Merchants were urged to dress up their windows to take advantage of the out-of-town visitors. Ads hawking every-thing from dry cleaning to car repairs, café meals to movies, took advantage of the aviation theme. Woodland Dairy came up with an "Armada Special" ice cream in honour of the flying racers, a blend of peaches and cream. School trustees were the only group to put on a dour face: school children would not be given a half holiday to watch the show.

Advance ticket sales were going like wildfire, according to the ads. A ticket got you a reserved parking spot around the field, and pedestrian access to the airplanes once they had all landed. The ads promised "50 Airplanes in Stunts and Thrills that will make you Gasp." According to the *Edmonton Journal*, over 35,000 people crammed the airfield to watch the show. A loudspeaker installed in the hangar kept everyone informed of the arrival of the planes, and then of the events.

A Ford Trimotor surrounded by admirers at the air show in Edmonton. Provincial Archives of Alberta, A5288

Five tank trucks from Imperial Oil stood ready to refuel the planes. Two fire trucks stood ready in case of disaster. The planes raced overhead and landed at terrific speeds, the pilots doing their best to get good times for the competition. The stunting, arranged by the aero club, included aerial moves and some air-to-ground action. Small planes chased a truck zigzagging across the landing field, trying to hit it with paper flour sacks. A parachutist from Calgary jumped from a plane at 600 metres. Three planes from the Edmonton club tried bombing a fort put in the middle of the field. The looping and stunting were spectacular, and the planes flew so low that their wakes blew hats off the heads of surprised spectators.

With the completion of the stunting, crowds surged onto the field to examine the planes and talk to the pilots. The flyers and their crew members were impressed with the interest and size of the Edmonton crowd. Later that evening, the Chamber of Commerce and the aero club hosted a banquet for them.

It was on to Calgary on the morning of 18 September. The planes rose into the sky at one-minute intervals, each trying to make the trip in record time. As in Edmonton, excitement had been building in the weeks before the tour's arrival. Early registration for parking earned a chance to win a novelty airplane smoking set.

A Great Civic Event

The National Air Tour, a major event in the aviation history of Calgary, Edmonton, and Lethbridge, gave residents the chance to see some of the newest and most technically advanced aircraft. The official directory produced for the Edmonton stop of the tour listed forty-eight planes competing for the trophy. Here are a few samples:

This is the new Ford high speed transport described as the fastest multi-motored plane in the world. It has a top speed of 155 miles [250 kilometres] an hour. In appearance, it will be one of the real sights of the tour; all of its exposed parts being burnished and polished until the plane shines like a jeweler's creation.

This is the famous efficient Bellanca, flown to victory recently in the National Air Races in Chicago. It will be piloted by J. W. Smith, Bellanca test pilot, and with its load-lifting ability, speed, and fine cabin fittings for passengers, will be a serious contender with any plane entered for tour honours.

This is the new all-purpose Curtiss-Wright multi-motored monoplane. This is an eight-place plane, particularly suitable for aerial photography and for use on airlines.

Programs selling for 15¢ provided the schedule as well as information about the pilots and their planes. Each program had a serial number which was entered for a draw; prizes included an aviator's watch and an airplane ride to Lethbridge.

The first plane, a Ford Trimotor, hurtled into Calgary after just one hour, eighteen minutes, and twenty-two seconds. In the next fifteen minutes, the remaining eighteen racers zoomed down from the sky. Thousands of people watched the spectacle, and swarmed onto the field once all the planes had landed. The Hudson's Bay Company chose this moment to light a giant neon beacon it had constructed in the heart of the city. The beacon was nineteen metres high, and its tremendous power came from forty-eight, three-metre neon tubes. The tubes threw a stationary light that was visible for 160 kilometres. "Faith in the Progress of Air Transportation" had spurred the beacon's construction, the Bay maintained. The official lighting ceremony featured Jock Palmer circling in his airplane, looking, the *Calgary Herald* reported, "like a flaring meteor as it soared and dipped, trailing brilliantly coloured lights from its wings and fuselage." Fireworks, an aerial armada from afar, and a giant neon welcome sign: flying had never been more magical.

Then it was Lethbridge's turn to host the flyers. A volunteer committee had been working hard to see that the tour's short stopover went smoothly. Each plane would be met by a "checker" as it landed, and guided to its spot on the airfield. A crew headed by Charles Elliott was in charge of servicing the planes with gasoline, oil, and grease. A lunch, catered by the Ladies Aid of Wesley United Church, would be held in the hangar for the flyers and local officials. Crowd control would be managed by the RCMP. The committee advertised throughout southern Alberta, and sent special invitations to the mayors of surrounding towns. Tickets selling for 25¢ got you into the parking area, and then onto the field once the planes had landed. The children of Lethbridge got the day off school to attend the event.

Around 11:00 AM on 19 September, the planes roared overhead and one by one touched down on the newly prepared landing area. They all landed safely, although one plane had hit a bird on the trip. Feathers flew and the plane went into a tailspin. The pilot had been flying at high altitude, which gave him enough room to get the plane righted. When all were down the crowd poured onto the field. The flyers praised the airport and the facilities. Elliott and his crew moved smartly to service the planes. A few stunts were performed, lunch was eaten, and then, shortly after 2:00 PM, the planes raced off for Great Falls, Montana.

The flyers and their mechanics had been impressed with the facilities they found in Alberta, and they were touched by the large crowds and eager welcomes awaiting them at each stop. Evidently they wanted to take a bit of the warmth of this Canadian reception home with them: customs officials removed about fifty quarts of beer, whisky, and sherry from the planes when they landed in Great Falls.

The RCAF responded to the increased interest in air tours as well. "On account of numerous requests for exhibition flying, a flight was put into intensive training at Camp Borden early in the Spring of 1931," stated its annual report. "This flight practised daily, weather permitting, on an average of four hours per day, and . . . formed the nucleus of the Trans-Canada Air Pageant Flight." One of the main attractions of the pageant was the RCAF Siskin Flight, a team of five planes. The air force also provided a Ford Trimotor and a Fairchild 71 to carry mechanics and officials for the tour. The tour was augmented by additional stunting planes, and performed from Charlottetown to Vancouver. In Alberta, it performed at Calgary and Edmonton, and stopped at Lethbridge.

The tour was organized by the flying clubs, the Aviation League of Canada, and the RCAF. Its broad aims were to advance aviation and stimulate interest in flying. It was hoped that the interest generated by the tour would spur airfield construction and a general improvement in facilities. Local clubs handled most of the arrangements, and the newspapers helped with publicity. The tour landed in Calgary on 17 July, and performed the next day. Cars lined the airport perimeter as the planes arrived, and one of the pilots shot fireworks from his plane, looping and banking "as a stream of brilliant fireworks fell like a waterfall toward the city below." A crowd estimated at between 25,000 and 30,000 attended the big show. Loops, rolls, upside-down flying, and the famed falling leaf stunt held the crowd breathless, and it was declared the best performance Calgary had ever seen. The Siskins were impressive, and the autogyro, a forerunner of the helicopter, was simply remarkable.

What turned out to be a comic act kept the spectators on the edge of their seats. A man was invited to look at one of the airplanes close up. He'd never seen one

The Ford tour stopped for refuelling at Lethbridge before heading off to Great Falls, Montana. Cars and crowds ringing the field greeted the flyers at every stop in Alberta. Courtesy City of Lethbridge Archives and Records Management, P19662107000

before, went the story. The man had a good look, then climbed in for a better one. Suddenly, the plane lurched into life, shot down the runway, and took off. Everyone could see the poor man waving his arms and calling for help. The organizers let it go on for a while before they announced over the loudspeakers that it was, of course, not a terrified innocent, but an experienced pilot at the controls.

The excitement was repeated in Edmonton. The planes were to arrive on 28 July, and the show was set for the next day. The program promised thrills galore, from acrobatics and formation flying to something called crazy flying. Tickets for the show went for 50¢ for adults, and 25¢ for children. Ticket-holders would be able to view the planes up close after they had all arrived on the 28th. And even though it was raining, a good crowd was on hand, peering through foggy windshields for the first glimpse of the pageant planes. The way in for the planes had been difficult. Heavy cloud and relentless rain had forced them to fly close together just thirty metres off the ground, pooling their lights in an attempt to improve visibility. Suddenly, they arrived over the airport, and landed to a symphony of cheers and honking horns. The Siskins, flying just three metres apart, were "glistening like silver darts," the *Journal* reported.

That night, the planes gave Edmontonians a hint of what was to come. The RCAF Siskins did formation flying and some individual stunting. Spins, loops, and dives at 370 kilometres per hour kept the crowd gasping. Barrel rolls and inverted flying had the same effect. Then, just after 9:30, a lone plane climbed into the air. At 900 metres the pilot started doing stunts with brilliant fireworks streaming from his plane. This stunning performance brought the evening to a close.

About 20,000 people gathered the following day, and enjoyed more than two hours of aerial thrills. The stunting was spectacular, with some comical acts in between. The "crazy flying" turned out to be a "country visitor" unfamiliar with the flying of a plane, or so the crowd was told. As the pilot skimmed over the field, wiggling his wings barely above the heads of the spectators, some of them must have wondered if, indeed, this wasn't the case. The Siskins were spectacular, diving and looping wing-tip-to-wing-tip before breaking up for individual acrobatics. The group then reunited for a perfect formation landing. Pilots from the Edmonton aero club bombed a "fort" in the middle of the field, and the parachute jump went off without a hitch.

The autogyro sparked the most curiosity. Described as a "whirling windmill contraption," its four rotary blades allowed it to go up and down in an almost straight line. It was one of the hits of the pageant, and interested crowds gathered around it whenever they were allowed on the field.

It was over much too soon. The organizing groups, the Gyro, Kinsmen, and aero clubs, expected to realize a healthy profit from the over 12,000 paid admissions. The Gyro club could continue to build playgrounds around the city, the Kinsmen would be able to further their work with crippled children, and the aero club hoped to get out of debt. Many people said that it was the kind of show they had wanted to see last year with the Ford tour; it had just taken Canadians to do it.

The planes stayed in Edmonton one more day, carrying passengers aloft. Then they left for Saskatchewan to continue the tour.

July 1931 was a month Edmontonians would not soon forget. An armada had swept out of the sky, then stayed to delight and thrill them. But just a couple of weeks earlier, two slight aviators climbing out of a single plane toward the end of a gloomy afternoon had caused nearly as much excitement. Wiley Post and Harold Gatty, dubbed "two gentlemen in a hurry," were flying around the world, from New York to New York, as fast as they could. On the home stretch, they had landed in Edmonton, where the aviation community was proud to host two such famous aviators.

They had left Fairbanks, Alaska, ten hours previously. Wind and rain had prolonged their flight, and the 2,000 to 3,000 people waiting patiently at the airport were becoming anxious. It had been raining all day in most of northern Alberta, reducing visibility and turning Blatchford Field to mud. Work crews had been trying to drain the puddles and fill the potholes with sand, but the rain quickly undid their work. About 4:30 PM the "Winnie Mae" dropped from the clouds and prepared to land. At 4:35 it came to a full stop on the field. The crowd surged toward it, taking no mind of the Edmonton ooze that quickly coated shoes and pant legs. The aviators emerged to cheers and out-thrust hands.

The cold, tired men were taken to the airport. When the buzzing in their ears

Although they were tired and weak when they landed in Edmonton, Wiley Post and Harold Gatty (in overalls) took time to speak to the press. Their progress was regularly reported around the world. Courtesy Provincial Archives of Alberta: Alfred Blyth Collection, Bl. 81

subsided, they answered questions from the press, and the news quickly travelled from Edmonton's hangar to radios across the country. NBC was also there, sending the story of the flight into the far reaches of the United States. Then it was back to work for Post and Gatty as they conferred with meteorologists and aviators about the weather and the next stage of their route.

Blatchford Field was too muddy to permit the Winnie Mae, with a full load of fuel, to take off. An inspired voice suggested, "Why not try Portage Avenue?" The aviators had a look at it and declared it the best runway they had seen on their whole trip. The plane was towed by tractor to the road, and a guard set up for the night. Wind and rain dictated a delay. Post and Gatty were taken to the Macdonald Hotel for a few hours' sleep. At 3:39 AM on Wednesday, 1 July, they lifted off from Portage Avenue on their way to Cleveland. A crowd was there to see them off, as usual, and the next news Edmonton heard was of the tumultuous welcome the two received when they arrived at New York.

A day or two later, Portage Avenue was again pressed into service. Reginald Robbins, an ambitious Texan, wanted to fly non-stop from Seattle to Tokyo. A Ford Trimotor piloted by Jimmy Mattern and Nick Greener was flying to Alaska where, fitted with extra gas tanks and special equipment, they would refuel Robbins's plane

Everything is ready for Wiley Post's arrival in July 1933. Even a movie camera has been set up. Courtesy Provincial Archives of Alberta: Alfred Blyth Collection, Bl. 137/2

in the air. Mattern and Greener stopped in Edmonton to prepare for the trip. Local aviators and meteorologists advised them on the best route north. Greener and Mattern equipped themselves with camping equipment, axes, food, and a rifle in case they were forced down.

A small crowd watched the takeoff from Portage Avenue at 3:30 Saturday morning, 4 July. After going off course over the mountains, they corrected their flight path and flew on to Fairbanks. This was the signal for Robbins to leave Seattle. He, too, made it to Fairbanks, but the story ended there. The wind was too strong to allow aerial refuelling, and Robbins's quest had to be abandoned.

It seemed flyers were hopping off all over the place. Newspapers were filled with reports of longer and longer non-stop flights. The Lindberghs were flying from New York to Tokyo via the Pacific. And almost as soon as Post and Gatty had established the world record for a round-the-world flight—nine days—other aviators were trying to break it. Long-distance flying was the order of the day. The news that airmen had landed in Turkey after a 8,000-kilometre, non-stop flight from New York, and that another globetrotting flight had landed in Berlin on its way to Moscow, pushed the Air Pageant off the front page of the *Edmonton Journal*.

Edmonton continued to be a stop-over for record-breaking flights. Blatchford Field and its manager, Jimmy Bell, had gained a deserved reputation for excellent service and topnotch organization. In 1933 Wiley Post was back in the Winnie Mae, this time circling the world by himself. He had sent instructions to Bell to have fuel and oil ready in anticipation of his arrival. Post hoped his stop would last about fifteen minutes. It turned out to be over an hour.

Thousands greeted the aviator as he pulled up to the concrete apron at the

A cheering crowd welcomed Post when he landed, and the service men sprung into action. Post stayed in Edmonton just eighty-seven minutes. Courtesy Provincial Archives of Alberta: Alfred Blyth Collection, Bl. 137/1

Edmonton airport at 6:14 AM on 22 July. The crowd had been gathering all night. Post, exhausted, was helped into the airport. Flying at high altitude over the mountains had left him with a headache. Cold compresses applied to his forehead as he lay on a couch in the radio room seemed to relieve some of the pain. There were quiet conferences about the weather, and he politely answered questions from the media. He accepted a cup of coffee and a sandwich, but refused to take a shower as recommended by the two doctors who were on hand to render medical assistance. Meanwhile, the Winnie Mae was being refuelled and having some spark plugs replaced, then it was towed by tractor once again to Portage Avenue. Eighty-seven minutes after he landed, Post roared off into the clouds, taking two thermoses of ice water, some sandwiches, and some apples and oranges. He landed safely again in New York, shattering the record he and Gatty had set two years before.

A smaller but no less genuine welcome awaited Jimmy Mattern when he limped into Edmonton a few days later. Mattern had begun his round-the-world flight before Post, but had gone down on the Siberian peninsula. Cruel air temperatures had stiffened his oil to the point where it would not move. The engine had overheated, and he was forced to land. Bitterly disappointed, he clung to the hope that he could get another plane and finish the trip, although he would not be able to break Post and Gatty's record. Still stuck at the remote station of Anadyr in Siberia when Post roared overhead, Mattern gamely relayed messages to him. When Mattern finally got to Edmonton on 27 July, he climbed onto the roof of the airport hangar and spoke to the crowd, thanking them for their generous welcome.

Not all sky riders were so daring or exotic. Many ordinary citizens were getting into the act. More planes were available from distributors and, as the 1930s got bleaker, more planes were available from commercial concerns that couldn't hold onto their business.

George Ross was a rancher in southeastern Alberta. A veteran of the Royal Flying Corps, Ross decided that he could oversee his vast herd better from the air. In 1929 he purchased a Curtiss Robin cabin plane. It gave him a marketing advantage, he explained to a reporter from the *Edmonton Journal*. He could fly to Calgary, Great Falls, and Saskatoon in search of better prices. His log book for the years 1929 to 1934 shows trips to various ranches, and frequent flights to cities and towns including Lethbridge, Lacombe, and Waterton. The flying rancher attracted quite a bit of attention. The *High River Times* reported on his novel way of doing business, and quoted Ross as saying that he carried a parachute with him at all times. There were so many eagles in southern Alberta, Ross explained, that if he hit one and damaged his plane he might need to use the parachute.

Dr Alexander Scott, who had come to Bassano to practise medicine in 1911, visited his patients by car and, in the winter, by horse and buggy. But Scott also had an abiding interest in aviation, and by the 1930s he had bought one plane, a de Havilland DH60G Gypsy Moth, and helped build another. He used local men as pilots, and flew across the countryside when medical assistance was required.

Some people were active in flying simply because they had always liked tinkering with things mechanical. Joseph Austin from Ranfurly, Alberta, owner of the Ranfurly

Garage, bought an airplane from the Eaglerock company in February 1930. He had earned his pilot's license at the Vegreville branch of the Edmonton and Northern Alberta Aero Club. He turned an old garage into a hangar, and was ready to go. For a few summers he barnstormed through the district, showing up at fairs or picnics and taking people up for a spin. On one flight, though, his plane grazed a haystack and suffered considerable damage in the ensuing crash. The plane was repaired, but the incident seems to have dimmed Austin's enthusiasm for flying.

W. A. "Red" Sherman of Granum was another aviator who flew for the sheer love of it. He took lessons from Great Western Airways in 1928, and bought a Gypsy Moth in 1931 from a bankrupt airline. He barnstormed, often with other pilots, for the next four years. His brother, Earl, was his mechanic.

By the early 1930s, there were landing fields from Grande Prairie to Vegreville to Coleman, and all the larger centres had well-equipped airfields. There was no shortage of people wanting to fly, either. Aero clubs and commercial companies crisscrossed the province offering rides. Although the Depression dictated that more and more bargain days were the norm, at a penny-a-pound, the citizens of Alberta did not seem to be losing their enthusiasm for flight. "Edmontonians Now Holidaying by Air," the *Edmonton Journal* declared in July 1930, and it was rare that a scheduled flight was without passengers.

The Alberta Provincial Police lobbied the government for an airplane in the late 1920s. It would be particularly useful, the commissioner argued, in search and rescue work and in patrolling the Crowsnest Pass and Montana border for liquor smugglers. The administration didn't agree. It felt a flying machine was an "expensive

Model airplane clubs were popular diversions for youth in the 1930s. Courtesy City of Edmonton Archives, EA–160–415

luxury," and the police were told to lease one on the few occasions one was needed. The RCAF used an airplane in 1931 for the world's first aerial buffalo census. Near Fort Fitzgerald, in Wood Buffalo Park, the herd was counted and photographed by air. The plane flew so low, the airmen reported, "on one occasion a bull buffalo turned and charged at the plane."

Model airplane clubs were doing a good business as well, and meets and contests were popular. "The Adventures of Captain Jimmy and his dog Scottie," a weekly newspaper column, followed the hero on aviation adventures in exotic locations.

Sky riders could take much of the credit for popularizing aviation. But for establishing the industry, for doing the work-a-day activities that made flying a business, the commercial companies were still carrying the burden.

Southern Dreams

In southern Alberta, commercial aviation companies depended on the same kinds of business that had sustained them through the late 1920s. Exhibition flying and stunting, student training, passenger service, and charter flights remained their bread and butter. As the Depression deepened, however, it became harder to make a living at it. Fewer and fewer people in drought-stricken southern Alberta had the spare change for an airplane ride.

One aspect of flying that assumed more importance for these companies was student training. In his memoirs, *And I Shall Fly*, Lewis Leigh recalls learning to fly in 1929–30 at Southern Alberta Air Lines in Lethbridge. It cost $30 an hour for dual instruction. The landing field was a complex of hollows and gophers, and the winds

---

"Air Adventures of Jimmie Allen" on CFGP

Condensed from the *Grande Prairie Herald*, 7 April 1938

Designed to develop initiative, ambition, and a spirit of fair play among Canadian boys and girls whose country has the finest tradition of success in the history of the youthful field of flying, the "Air Adventures of Jimmie Allen," written by real aviators about a real boy, will be presented to youthful and adult listeners.

Already recognized as the biggest boy and girl program on the continent, with 4,000,000 members in its "Jimmie Allen Flying Clubs" and a series of Air Races by model aircraft built by the members, the series of broadcasts tell the story of 17-year old Jimmie Allen who worked at an airport, learned to fly and ran into a string of thrilling, wholesome adventures.

Real lessons in flying, which every boy and girl can learn, are heard over the air as "Speed" Robertson, a real flier attached to one of the biggest airports in the world, tells his young pal Jimmie how planes are flown, what the various instruments are for and how problems of the air are met by pilots.

that scoured Lethbridge meant that most instruction had to take place before dawn. Before students could be tested for their license, each had to put in a specified number of solo flying hours. Not surprisingly, many potential flyers did not have the money to rent a plane for those hours, and so earned their time aloft by helping around the company's operation. Leigh went with Southern's pilots on barnstorming tours, selling tickets and assisting passengers into the planes. In return, he got to fly the aircraft. The precious hours he earned were spent practising the rolls, loops, and spins that were a staple of a barnstormer's livelihood.

Tweed Air Service, operating out of Lethbridge intermittently between 1932 and 1937, counted on student training to bring in revenue. Charlie Tweed travelled regularly to the Crowsnest Pass to instruct members of that flying club in 1935. Students from Blairmore, Coleman, and Bellevue were eager to take to the windy skies of the Pass. Most flying businesses of the time couldn't boast Tweed's five-year survival record. Small companies seemed to pop up one year, only to change form the next. Planes were sold, traded, or leased, and personnel moved about easily and frequently.

Ernie Boffa's career exemplifies the resourcefulness and determination aviators had to have to survive in southern Alberta in the 1930s. Boffa arrived in Lethbridge from Great Falls, Montana, in 1930, bringing a Waco that he leased to Southern Alberta Air Lines while he got his Canadian pilot's license. He also spent a good deal of time repairing and rebuilding damaged aircraft for other companies. By 1931 he had another Waco and took up barnstorming seriously, travelling with the All-Canadian Tour. He settled in Medicine Hat and joined up with Lewis Leigh, who had been let go from Southern. They decided to increase their earning potential by adding daredevil stunts to their routine. Their first one involved Boffa climbing onto the

Barnstormers blow into the Crowsnest Pass. Courtesy Glenbow Archives, Alberta, NA–4880–6

wing, wearing a leather harness attached to a parachute that had been tied to a wing strut. Boffa would dive off the wing, seemingly without a parachute, but then his rope would pull the parachute from the strut, to the great relief of the crowd.

This wasn't enough for Boffa, so he and Leigh devised another, more dangerous stunt. To the crowd below it would appear that Boffa was hanging by his teeth from a rope tied to the undercarriage. In reality, Boffa would be holding onto a knot in the rope. As the pilot, Leigh couldn't see Boffa once he had clambered down to the undercarriage, and had no idea how the stunt was going. A good crowd showed up for their first attempt. Leigh circled the field after Boffa had disappeared over the wing, and was horrified to see friends below trying to signal that something was wrong. Leigh, of course, had no idea what it was. He managed to land the aircraft, expecting to find his partner crushed under the plane. Boffa survived, little the worse for wear. Leigh had landed the plane carefully, and Boffa had managed to run along behind it. Leigh was reprimanded by the Civil Aviation Division of the Department of National Defence, and the stunt was not performed again.

Boffa decided to confine himself to piloting, and joined forces with parachute jumper Roy Lomheim. Lomheim did a double parachute drop: after his first chute failed to open he continued plummeting earthward. Then, at about 150 metres, he opened the second chute and floated safely down. There were usually a number of people wanting a ride in Boffa's plane after the performance.

Boffa offered charter flights as well. He delivered supplies to oil well sites and

Charlie Elliott, shown here with his Gypsy Moth at Taber in the late 1920s, was active in aviation in southern Alberta until 1932. The plane's wings folded back to make storing it easier. Courtesy Glenbow Archives, Calgary, NA–3277-67

took salesmen or businessmen on to their next calls. In 1935 he purchased an Eagle aircraft from an aviation company that had gone bankrupt. Later that year he traded the Eagle to a company that needed a larger plane, taking a Puss Moth in exchange. He now had a plane with an enclosed cabin, and could offer charter service year-round. He was able to scrape a living out of flying in the 1930s by involving himself in all aspects of the business, from repair and rebuilding to all types of flying. In mid-1936, he left southern Alberta for Saskatchewan to take up work as a bush pilot.

The weather, as well as hard times, influenced aviation developments in the 1930s. In the winter of 1937, for example, blizzard after blizzard buried southern Alberta, and ski-planes provided emergency transportation over blocked roads and railways. And of course the wind always affected flying. Lewis Leigh recalled how difficult barnstorming could be with little Moths in windy conditions. Landing was difficult enough on unknown fields, but wind often made it treacherous. Helpers would fan out onto the field, and the pilot would try to land as close to them as possible. When the plane touched down, the helpers would grab the wing tips and hold on until the aircraft was tied down: otherwise, the wind would toss it onto its back. Another barnstormer's trick was to head for a haystack if wind threatened to upset the landing; it caused less damage than being flipped over.

Developments out of Calgary were sporadic as well. Airmail cancellation was a great blow to the city. The deepening Depression was another factor. Calgary companies relied on charter flights, exhibition flying, student instruction, and passenger service to eke out a living.

In late 1931 Western Flying Service tried to fill the gap created by the collapse of

A plane being rebuilt at SAIT, with the proud class posing before it, in 1932. Courtesy Glenbow Archives, Calgary, NA–4669–4

Rutledge Air Service and Great Western Airways. Jock Palmer was instrumental in putting the company together. Two planes were purchased, one each from Rutledge and Great Western. By the spring of 1932, the company was operating from the Calgary airport. By the following summer, however, the company was no longer in business.

McLaren Air Service had a modest beginning in 1932 with a single Curtiss Robin plane. The company hauled fish under contract in the winter of 1932–33, and did some barnstorming in 1933 as well. Western Flying Service and McLaren tried to reverse their sagging fortunes by sponsoring a "Monster Air Pageant" in Calgary in July 1933. The pageant promised fun and frolic both in the air and on the ground, as motorcycles and cars were added to the stunting roster. On the big day, Freddie McCall took to the air for some spectacular stunting. But the afternoon belonged to Howard Sandgathe of Bassano who took home first prize in the balloon-bursting, bombing, and spot-landing competitions. Airplane relay races, formation flying, and parachute drops, including one by Vi Palmer, a local woman, rounded out the thrills. Many spectators attended the show but, as the *Calgary Herald* pointed out, the entertainment could be seen from outside the grounds, and the paid attendance was disappointing. By 1934, McLaren Air Service's single plane had been sent to British Columbia where one of the original partners in the company had located.

Chinook Flying Service was another company that tried to get into the business on the back of the defunct Great Western. It purchased three of the company's planes in early 1932, bringing to four the number of aircraft Chinook could put in the sky. Passenger carrying and barnstorming got them through the summer. One plane was based at Sylvan Lake and took tourists up for leisurely flights over the water. It did well, but there was little other business, and by fall the company could not meet its payments. The planes were repossessed, and Chinook ceased operations.

Aviation in the south remained on shaky ground until the late 1930s. Skyways, which began operations in Calgary in the spring of 1936, was another venture in which Jock Palmer was involved. The company began with a Command-Aire 3C-3, owned by one of the other principals in the company. For a year, the company survived on pilot training, charter flights, and passenger service. A company ad listed its fees, including $150 to study for a private pilot's license and $3 for a trip over the city. When the plane's engine wore out, however, the company was unable to replace it. Skyways faded out of existence.

Prairie Airways, founded in Moose Jaw in 1935, opened a base in Calgary in 1937 with a Cessna C-34 Airmaster. The new company ran a charter service, flying businessmen out to their companies' oil wells. One customer was so impressed with air travel that he bought the plane. Prairie Airways shut its Calgary office.

The new company, Midland Airways, was formed by oil men who believed in the value of air transportation. But they hoped to realize a profit from pleasure trips as well. An advertisement in the *Calgary Herald* asked "Have you seen Calgary from the air?. . . or flown over Canada's Greatest Oil Field–Turner Valley?" A Saturday trip for three over the oil field could be had for $4.95 per person. The plane was damaged in a rough landing in 1939, and the company didn't replace it.

Another company operating out of Calgary in the late 1930s reached for a piece of the oil industry's aerial business. Northwest Air Service Limited, begun in the fall of 1938 by a couple of oil companies, had ambitious plans. With a new Beechcraft D17-S biplane with retractable landing gear, and calling itself "Canada's Fastest Charter Service," Northwest announced that its "deluxe air-conditioned cabin plane" was ready to charter "from any points desired." The Beechcraft was also equipped for night flying, and had a two-way radio. This company lasted about a year, flying in Alberta and the United States, before the plane was sold to another company in the province.

Private aviation companies faced the loss of charter revenue from the very businesses that might have hired them; many of their potential customers were establishing their own aviation sections. Imperial Oil and Shell, for example, regularly flew their own planes into the province, and local and regional companies were adding flying to their lists of transportation options. Canadian Western Natural Gas, Light, Heat and Power Company Limited, established in Calgary in 1911, was one of them. In 1936, the company bought a Stinson Reliant monoplane to transport employees and help oversee its scattered wells and pipelines. The plane was particularly useful in January and February 1937, when the blizzards that had kept Tweed Air Service hopping throughout the Lethbridge district made it impossible for ground transportation to reach many well sites. Crews and equipment were flown down in the Stinson.

Consolidated Mining and Smelting Company of Trail, British Columbia, was another company that used its own aviation service for exploration and personnel transport. In 1931 it opened a base at the Calgary airport, determined to use airplanes to sell its brand of fertilizer throughout Alberta. The *Grande Prairie Herald* carried a report on 25 March 1932 headlined "Farmers Experience Thrill of Being Visited by Airplane." A salesman for the company had been touring the district by plane, the pilot setting the machine down at various farms. The salesman reported that sales were going well here, and his next stop was the Peace River country. The *High River Times* and the *Lethbridge Herald* also reported on Consolidated's flying fertilizer sales technique, indicating that the company blanketed the province from the air. Its Calgary base remained open until 1935.

For private companies, however, there simply wasn't enough business in southern Alberta during the 1930s. Charter opportunities, such as the time Charles Tweed took a wool buyer on a tour of sheep farms, were either one-time propositions or emergencies, and not the sort of thing to develop into regular scheduled flights. Flying didn't stop, though. Companies evolved and names changed as determined flyers squeezed another year's flying out of a few students and some starry-eyed fairgoers who wanted to see their community from the air.

From Sea to Sea to Sea

O n 12 May 1930, in the elegant Macdonald Hotel overlooking the North Saskatchewan River from downtown Edmonton, a group of people sat down to dinner.
They were in a celebratory mood. They had come to honour a team of world champions, and two men who were pushing the horizon further and further away. The
happy gathering feasted on "Punch" Rolls, Roast Lamb "McBurney" Style, "Mae
Brown" Potatoes, and Coupe "Wop" May.

The champions were the Edmonton Grads, officially known as the Commercial
Graduates Basketball Club. Since 1915 they had dominated women's basketball, win-
ning several national and international titles. They had just won another Underwood
International Trophy, defeating the Chicago Taylor-Trunks in a thrilling series. The
Grads, including Mae Brown and Margaret McBurney, were the toast of Edmonton.
The two men being honoured were C. H. "Punch" Dickins and Wilfrid Reid "Wop"
May, recipients of the McKee Trophy. The trophy was awarded to the individual who
had made a significant contribution to the development of aviation in Canada in a
given year. Dickins had won the honour for 1928, May for 1929. Dickins had flown
almost 6,400 kilometres through northern Canada, carrying personnel from an exploration company on an aerial survey of the west coast of Hudson Bay. May's achievements included the humanitarian flight north with diphtheria anti-toxin, and establishing a mail route down the Mackenzie River to Aklavik.

After the banquet, a parade wound its way through Edmonton to the legislature.
The Newsboys' Band, the 49th Battalion Band, and the Canadian National Railways
Pipe Band provided musical accompaniment. The guests of honour rode in cars, as
did civic and provincial politicians. Boy scout troops and girls from basketball teams
throughout the city marched proudly along. Cheering crowds lined the streets. More
people waited at the legislature to greet their heroes: a "Sea of Humanity," estimated
by the *Edmonton Journal* to number between 25,000 and 30,000 people. After the
speeches, each of the Grads was presented with a case containing fifty pieces of
silverware, and a travelling wardrobe bag. Dickins received a "handsomely engraved
wristwatch," according to the *Journal*, May a "beautiful clock."

The next day, it was back to work for everyone. The Grads returned to their jobs and evening basketball practice. For Dickins and May, the ice in the northern rivers was breaking up; the in-between season would soon be over.

Down North

The *Edmonton Journal* reported in May 1930 that, during the previous winter flying season, Western Canada Airways and Commercial Airways had logged more than 200,000 kilometres in the north, carrying 780 passengers and over 50,000 kilograms of express and mail. Planes were now massed at Fort McMurray, ready to start the summer season.

The two flying seasons were separated by breakup in the spring and freeze-up in the fall. These were the in-between seasons when planes couldn't land. The rivers and lakes were the runways and hangar aprons of the north. Float-planes needed a clear stretch of open water, ski-planes a flat stretch of snow-covered ice. Winter or summer, water was the key to northern transportation, and the high-flying technology had to wait until the ancient waterways were prepared to carry it once again.

The aviation companies began to advertise as the in-between seasons drew to a close. On 23 May 1929 and again on 13 June, the *Financial Times* carried an ad from Western Canada Airways. The first announced that limited air service was now available from most of its bases, including Waterways, Alberta, down the Mackenzie. The second advised that summer air schedules were now in operation from all bases.

Regular northern flying had begun in the late 1920s. As manufacturing expanded across Canada, the demand for raw materials grew. Many looked to the Canadian Shield, a stony quilt flung across the north, to supply the demand for metals and ores. Geologists, mining engineers, and prospectors spilled from railheads into ships

Fort McMurray was a base for a number of planes that regularly flew down north.
Courtesy University of Alberta Archives, E. Burton Collection, 83–140–26

and barges, then continued in canoes or on foot, scouring the rock. Some worked for the federal government, others for mining companies, many just for themselves. From Quebec to Ontario, across the west and into the Northwest Territories, they searched for their treasure throughout the short northern summer. But too many days were spent just getting there. Long journeys down rivers and across portages, then hours spent hiking over difficult terrain ate into precious prospecting time. It wasn't long before the airplane came into general use, turning weeks of travelling time into days, and days to hours. Prospectors could be dropped off where they wanted to be, then picked up a week or two later, or as winter threatened. But the airplane did not operate in isolation. A well-integrated system of water transportation ensured that fuel caches and supplies awaited pilots and prospectors when and where they were needed.

Several businesses served only prospectors and the companies they represented. The Northern Aerial Minerals Exploration Company (NAME), formed in Toronto in 1928, specialized in transporting mining personnel and equipment into the north. Edmonton became the headquarters of their Alberta and Northwest Territories operations. It was while flying for a similar company, Dominion Explorers Limited, that Punch Dickins made his 6,400-kilometre odyssey across the Arctic.

Time was not the only factor in making the airplane ideal for northern operations. Aerial mapping and photography were by then well developed. Photographs taken from the air could be examined at company headquarters, and prospectors could then be sent to the areas that looked most promising. Planes suited to hauling and harsh conditions had made winter flying almost as common as summer flying. Companies could choose from Fairchilds, Bellancas, Noorduyn Norsemans, Fokkers, and Junkers, all large enclosed planes capable of hauling heavy loads on long flights. The pilots and engineers were better dressed, too. A picture of NAME staff taken at Winnipeg in 1928 shows them in fur-lined coats and thick parkas. Ernie Boffa, still flying an open-cockpit Moth in northern Saskatchewan in the early 1930s, wore duffel coats, parkas, felt-lined mukluks, and caribou socks sewn by local Native women.

As engines and struts and skis and anything else that could break or malfunction in the cold did, engineers and pilots got better at fixing them and devising new ways to get the equipment to do what it had to. The case of Harold "Doc" Oaks and the Elliott brothers is a good example. Oaks was a pilot with NAME and Western Canada Airways. Carmen and Warner Elliott operated a thriving business making sleighs, toboggans, and boats at Sioux Lookout, Ontario. Unhappy with the skis on WCA planes, Oaks devised a better one and persuaded the Elliotts to build it. With Oaks's encouragement, they began manufacturing aircraft skis, and soon became a prominent supplier.

In northern Alberta, flying was becoming more common. Commercial Airways' regular run down the Mackenzie was an important development, as was the increasing number of communities served by airmail flights. More and more people, from nurses to priests, bureaucrats to salesmen, and trappers to traders, were at least occasionally using planes for some of their business. Flying trappers in to their trap lines for the winter, and out again in the spring was a particular challenge because of their dogs, a snarling, unpredictable, and often vicious cargo.

It was difficult, irregular work. In a report sent to Western Canada Airways' headquarters in 1931, Punch Dickins explained the difficulties in collecting on personal accounts in the north. With no banks and little money in circulation—and with the fur trading companies wanting to keep it that way—it was often impossible to get accounts settled. "An example of collections received is given when on my last trip," Dickins wrote, "I arrived back at McMurray with $15.00 in cash, $240 in 'Wolf Bounties,' $320 in beaver and destitute ration orders, 400 muskrats, 2 red foxes, 2 cross foxes, 7 marten, 5 mink, 1 lynx and 50 extra muskrat to come and go on."

Local commercial operations also created a role for the airplane. Winter fishing was a dependable source of income for people in northern Alberta and Saskatchewan. The fish were hauled by horse-drawn sleighs to railheads, then shipped to larger centres for distribution. Transporting the fish to the railroad was hard work, made miserable by frigid temperatures and lashing winds. Enterprising aviators

Geologists and mining engineers used planes in their field work. In 1930 the University of Alberta's A. E. Cameron went on an extensive tour of the north. Commercial Airways' Bellanca Pacemaker CF–AIA can be seen here on a northern lake behind the geologist's camp. Courtesy University of Alberta Archives, E. Burton Collection, 69–160–260

stepped in to make this part of the journey faster and easier.

A glance at the *Northern Gazette* of Peace River for 1932 and 1933 shows the W. J. Menzies Fish Company hiring planes to fly the catch from local lakes to Faust where it could be transferred to rail cars. The company expected the aerial method would actually be cheaper than using sleighs. Several air companies freighted from lakes in the area, including Whitefish, Wabasca, Peerless, and Lesser Slave. Faust, the paper boasted, was "quite an airplane base." Similar activity was taking place at a number of communities across northern Alberta.

For Alberta, though, the real meaning of down north didn't come until the explosion of mining development around Great Bear Lake, and later Lake Athabasca and Yellowknife. When a prospector named Gilbert Labine got a pilot named Leigh Brintnell to drop him off at the eastern end of Great Bear Lake in August 1929, it was the beginning of a new era for aviation in Alberta. Labine, looking for copper deposits, staked thirty-three claims. Soon after Punch Dickins picked him up for the trip out, Labine noticed discolouration of the rocks below. He got the location from Dickins, and made plans to return the following year.

In March 1930, he found what he had suspected: rich deposits of silver and pitchblende, the source of radium and uranium. More prospectors were combing the area as a result of his previous year's copper discoveries, but Labine was the only one to recognize pitchblende. He quickly staked ten claims. When word of his latest discovery leaked out, the rush was on.

Great Bear Lake was over 1,200 kilometres north of Fort McMurray—1,200 kilometres of rock and trees and water. Fort McMurray became the jumping off point, where the railway ended and the highway in the sky began. Aviation companies already had bases there, with docks, slipways, hangars, and fuel depots. Imperial Oil had established fuel tanks at Waterways in 1922 to supply the river boats, missions, police, and Hudson's Bay Company posts. With the concurrent growth of mining and aviation, the depot expanded to serve its new customers. By 1932, several thousand claims had been staked in the Great Bear Lake area, and the rush to get personnel and materials there opened up new opportunities for flying companies.

Since 1930, several aviation companies had been vying for a share of northern business. Western Canada Airways soon absorbed Commercial Airways and became one of the major players. Smaller interests would come and go, eventually leaving most of the field to the big companies. Spence–McDonough Air Transport Limited,

Down North

This was the term everyone from pilots to newspaper reporters used to denote travelling north from Edmonton. "Going Down North," for example, was the title of an article in the December 1938 issue of the *Nor'West Miner* that talked about the various means of travelling north.

for example, was begun by two pilots with a Fokker Universal, who began flying north from Edmonton in 1930. In early 1933, Canadian Airways Limited absorbed them, too.

But even at the height of the rush, 1930–1932, business was not unlimited. The cancellation of airmail contracts had forced companies to let their pilots, mechanics, and support staff go. The RCAF was releasing people, too, as the Depression strengthened its grip on Canada. Smaller companies sputtered and died as business dried up. Even though flying down north was picking up, it was not enough to absorb all the casualties of the general downturn in aviation.

The example of Lewis Leigh and George Silke is typical. Leigh, a former RCAF pilot, arrived in Edmonton from Sydney harbour, Nova Scotia, in 1932. He and Silke, another pilot, hoped to establish a western branch of Explorer's Air Transport. Seeing the shortage of business, Silke decided to try his luck in British Columbia, while Leigh tried to keep a second plane busy in Alberta.

As they had arrived in the summer, Leigh based his plane at Cooking Lake and barnstormed at nearby resorts. He and his wife, Lin, often slept in the plane on coils of rope, using the canvas engine cover for a blanket. Leigh got several contracts ferrying loads, and flying prospectors and their gear to Fort McMurray. In his own words, he was "manager, pilot, engineer, loader and unloader of aircraft." He was not paid a salary but a percentage of revenue, and that didn't amount to much.

One of the biggest problems was fuel. Every company had its own supply at various points throughout the north, and they were jealously guarded. Canadian Airways, for example, wrote to Mackenzie Air Service in 1933: "This is to advise you officially that effective this date, legal action will be taken against Mackenzie Air Service Limited, should they at any time use gasoline or oil from our caches, or supply dumps, without having written authority there from our general office at Winnipeg, signed by the Assistant Controller." Clearly, if you didn't have your own supply, you didn't fly. It was a slow and expensive business to bring fuel into the north. Most of it was transported by river barge, in barrels, and transportation was limited to the summer season. It often took a year or more to establish fuel reserves for the entire run. You couldn't accept flying assignments without fuel stores in place, but without flying assignments, there was no money to establish a fuel supply. Leigh managed to get gasoline on credit; Explorer's Air Transport signed for some, and he signed for the rest himself. He could now begin establishing fuel caches. In January 1933, he persuaded a friend to act as business manager so that he wouldn't miss potential contracts while he was flying or working in the plane. Leigh set him up in an Edmonton hotel in a room that doubled as an office. With fuel caches up to Fort Rae, they felt they were ready for the challenge of the north, and a flight to Great Bear Lake.

In February, Leigh carried a trader and a full load of goods up to Great Bear, flying over a period of days in bitterly cold weather. The trip was successful, though, and he managed to get a partial load for the flight home. For the next while he was busy flying supplies into Hudson's Bay Company posts and taking bales of furs out. There was little profit in the contract, but the work was steady. Then he was grounded by the Civil Aviation Department over a problem with his propeller. With the operation earning so little, the company decided to close down the business.

James Richardson of Canadian Airways knew the importance of flying down north. In the economic climate of the early 1930s, the run to Great Bear Lake and down the Mackenzie River was one of the few business ventures in Canada with the potential to expand. Punch Dickins, the area manager, now had six planes on these runs, compared to one in 1929. A 1933 ad for Canadian's northern operations referred to the company as the pioneers of the Mackenzie River district, and boasted four years' successful operations. With six modern aircraft, experienced pilots and licensed engineers, the ad went on, Canadian Airways could accommodate surveys, exploration, photography, transport, and airmail. Hauling for the mining developments at Great Bear Lake provided the bulk of the business. On 4 March 1932, for example, five loaded planes headed to Echo Bay on Great Bear Lake, returning on schedule two days later. Within a week, the five again went north, again fully loaded.

Day in and day out, the flights continued, but profits were still not large. Canadian estimated that its hourly costs on northern operations in 1933 ran to $43.75 while revenue came in at $33.16. The airline reduced its other operations to keep the northern work afloat. One method of increasing revenue relative to cost was overloading the planes. It was a practice Canadian followed only until 1932, according to Ken Molson, the author of a study of the company. Canadian charged that smaller companies regularly overloaded their planes throughout the decade to make flying profitable. Overloading was, of course, against civil aviation regulations. A plane has an approved gross weight at which it is authorized to fly. At this weight, it will take off in a certain distance, rise at a certain rate, and be able to withstand reasonable gusts of wind. Overloaded, the distance required to take off increases, the rate of

A Canadian Airways Junkers being loaded with goods from the north. Courtesy Provincial Archives of Alberta: Alfred Blyth Collection, Bl. 278/9

climb slows, and the plane's stability in the air is compromised. These conditions put stress on the planes and their engines.

It was difficult to enforce load restrictions in the north; in remote locations, federal officials were not on hand to carry out regular inspections. The practice of overloading caused friction between companies, as those that abided by the regulations sometimes reported those that did not. It seems to have been a tactic for reducing or eliminating competition. It was one of Lewis Leigh's competitors, for example, who reported that he was flying with a laminated wood propeller instead of the metal propeller required by civil aviation regulations, and it was this that got him suspended.

Canadian wasn't the only substantial carrier running into the north from Alberta. In December 1931, Leigh Brintnell established Mackenzie Air Service, with offices at Fort Smith and Edmonton, and headquarters at Edmonton's Municipal Airport. Brintnell had been a pilot with the Ontario Provincial Air Service, but left in 1927 to fly for Western Canada Airways. He quickly rose to the position of Assistant General Manager, and when amalgamation in 1930 produced Canadian Airways, he became Assistant General Manager for its western lines.

Brintnell left Canadian in 1931 to establish his own company. With backing from

Matt Berry, a well-known pilot, showing a common method of transporting canoes. Courtesy Provincial Archives of Alberta, 66.122/116 (24)

Anthony Fokker, Brintnell bought two Fokker Super-Universal aircraft. He already had contracts to transport ore from the Great Bear Lake mine of Gilbert Labine's Eldorado Mining and Exploration Company. The two planes, painted in Mackenzie's gold and green corporate colours, were soon hauling supplies and equipment into Great Bear and bringing ore out. Brintnell's business grew substantially over the next few years, and he added Bellanca, Fairchild, Barkley-Grow, and Noorduyn Norseman aircraft to the fleet.

A 1935 advertisement for the company boasted of a modern fleet giving service in the Mackenzie district, to Great Bear Lake, to northern British Columbia and the Yukon, and to Lake Athabasca. The latter had been added to the mineral map in 1934 with the discovery of gold deposits there. As prospecting and development got under way, it became another aerial destination from Fort McMurray. There were mining developments at Yellowknife, too, as the gold that had been discovered there thirty-five years ago during the Klondike rush was finally brought into production.

It is clear that much of this development would not have occurred as quickly without the airplane, but it is also clear that the airplane fitted smoothly into a system of transportation that was by no means abandoned when regular flights began. The Mackenzie River remained the highway to Great Bear Lake. The route was open from the end of July to the beginning of October, and heavy freighting was done with steamers and scows. The twenty-six-kilometre portage between Fitzgerald and Fort Smith was served by a graded gravel road. In fact, in the earliest days, only the urgent and the indispensable items were flown in, as freighting by air cost $3.30 a kilogram. Bulky or heavy cargo and much of the fuel supply continued to be brought in by boat.

Mining development at Lake Athabasca showed a similar integration of transportation systems. Most of the equipment was shipped by rail to Fort McMurray, then transferred onto river carriers. Mine cars, steel tracks, diesel engines, timber, mill machinery, a hoist, tractors, boilers, and supplies and food for mine workers for one year all travelled downstream to the mine site. The mine developer, meanwhile, whizzed back and forth between McMurray and Lake Athabasca in the air, overseeing preparations to receive the equipment. The two methods of transportation were complementary, and no time was lost. When they were ready to install the equipment, it was waiting at the dock.

The dimensions of some mining equipment dictated that it be transported by boat, a fact noted in the March 1935 *Canadian Mining Journal*. The manufacturing industry was responding to this constraint. The *Journal* noted that more and more machinery was being sectioned by the manufacturer to make loading it into planes possible. Still, the day-to-day ferrying of supplies and people remained the plane's strength: one of Mackenzie Air Service first deliveries brought fresh vegetables to the dining room at Eldorado's winter camp.

One of the most traditional of Canadian water transportation methods, the canoe, remained crucial in northern mining development, particularly for the prospector. It also provided a challenge to flyers. How do you transport a canoe by air? It was a question deemed worthy of study by the National Research Council. In the end, the best method was to lash it to the float struts.

Along with Mackenzie Air Service and Canadian Airways Limited, Grant McConachie had aspirations to dominate northern flying, although he didn't just depend on the mining companies for business. The flamboyant McConachie was ready to put together a deal to do just about anything. He started flying in late 1931 with a used Fokker Universal he had purchased with a loan from an uncle. His first contracts included flying for Professor Rowan's migration experiments with crows and hauling fish from Cold Lake to the railway station at Bonnyville. McConachie also carried groceries and other supplies into the snow-bound communities in the Cold Lake area on his return flights. Sometimes he took passengers as well, and they must have reeked of fish as they staggered out of the fragrant aircraft at journey's end.

From his fish-hauling contract, McConachie was able to pay his and his mechanic's living expenses in Bonnyville, and repay the loan from his uncle. But he hadn't taken into account the wear and tear over 600 hours of flying had done to the plane. The Fokker could not be declared airworthy without a complete overhaul, something McConachie could not afford. Again, he borrowed money from his uncle.

He spent the summer of 1932 barnstorming through central Alberta, and that fall he secured another fish-hauling contract, this time from a lake in Saskatchewan to the railway at Cheecham just south of Fort McMurray. He had also found another

Canoes were important both in the search for minerals and for unloading float planes. Courtesy Provincial Archives of Alberta, A5287

investor who had a Fokker and a Puss Moth, and Independent Airways was born. It was not an auspicious beginning, however, as McConachie crashed one of the Fokkers during a takeoff from Edmonton in November. "As soon as the wing tip hit the ground," he recalled, "the plane started to cartwheel, from wing tip to nose, to the other wing tip, to the tail, to wing tip, to nose. . . . For two city blocks, the Fokker jolted, spun, whirled and crashed, rolling itself up into a ball of twisted metal and tattered fabric." The plane was a wreck. McConachie, with multiple fractures to his legs, a broken hip, broken fingers and ribs, and a shattered kneecap, was not much better off. Undaunted, he was back in the air by spring.

Professor William Rowan of the University of Alberta studied the migratory habits of birds by airlifting crows to various parts of the province and releasing them. Rowan had the birds' tails dyed yellow so they could be spotted easily. He encouraged Albertans to shoot any they saw and express them back to him giving the location of the sighting. By tracking the crows' movements, Rowan could tell when migration began, and pinpoint the changes in sunlight that triggered the migratory response. In this photograph, pilot Grant McConachie gets ready to load crates of crows into his plane. Courtesy University of Alberta Archives, William Rowan Collection, 69–16–2250

He continued his hauling contracts, and expanded into contract flying in northern British Columbia. His company went through a couple of metamorphoses, emerging first as United Air Transport, then as Yukon Southern Air Transport in 1939. That was the year he managed to get hold of a Ford Trimotor and coasted into Edmonton with the plane most people hadn't seen since the big air shows of five and six years before. He added the Trimotor to his fleet, hauling fish from Peter Pond Lake in Saskatchewan to Cheecham, Alberta. With four planes, four pilots, and four mechanics working six days a week, McConachie hauled half a million kilograms of fish that winter of 1934–35.

McConachie did not have his gaze fixed solely on northern flying. He barnstormed with the Ford, too, lumbering into prairie towns to take the locals up for a spin. The Trimotor also made the first commercial crossing of the Rocky Mountains. When a businessman needed to get from Calgary to Vancouver in a hurry, and was looking for a big plane to take him there, McConachie jumped in with the Ford. It was a chance to get some publicity, and make a tidy profit as well. A large crowd was on hand to see them off from Calgary on 16 May 1935 at 9:30 AM. Ice on the wings forced a landing in Grand Forks, where they refuelled, but McConachie could not get back into the air until a civic luncheon had been held to honour the unexpected guests. This delayed the flight considerably, and Vancouver citizens waiting for the conclusion of the flight became very anxious. Once safely on the ground in Vancouver, the pilot and his passengers were entertained at a banquet, given under the mistaken assumption that this had been the inaugural flight of a cross-Rockies service.

The Ford was not equipped with floats, so it couldn't land on water. This made it of limited use for northern flying, and McConachie managed to trade it to a Yukon company for a Fairchild float-plane. The flight up to deliver the plane, and the cross-country test flights the prospective buyer insisted on, convinced McConachie that his own future in aviation lay in the Yukon.

McConachie's experience is illustrative of the vagaries of the flying business in the 1930s in northern Alberta. He was often short of cash. He relied on used planes to expand his fleet. He was continually in search of contracts and operating capital. But he and his pilots demonstrated the same determination to wring a living out of the skies that characterized Canadian Airways and Mackenzie's operations. In 1936, he looked to mail contracts for a source of regular winter revenue, and over the next couple of years inaugurated runs from Edmonton to Whitehorse, and between communities in British Columbia. He also began regular flights to Peace River and Grande Prairie. By this time Mackenzie was hauling mail in northern Alberta and beyond, and Canadian Airways was delivering from Fort McMurray to communities on the Athabasca, Slave, and Mackenzie rivers, and to Athabasca, Great Slave, and Great Bear lakes.

So much activity was heading down north that the western office of the Controller of Civil Aviation moved from Regina to Edmonton in 1934. Haulage kept steady right through the 1930s. In 1935, Mackenzie Air Service hauled several diamond drill compressors to Lake Athabasca, as well as a neatly disassembled flotation mill that weighed 9,000 kilograms to Great Bear Lake. In 1937 Russ McLaren, who had tried to make it on his own with McLaren Air Service, moved to Edmonton to join Canadian Airways as an engineer. In 1938 he was routinely flying 1,600 kilometres a day, six days a week. Sometimes they flew on the seventh day to make up lost time. McLaren remembers as many as four flights a day into Yellowknife, with each aircraft in the air eight or nine hours daily. "In one two-day period," he recalled, "we moved thirty-five passengers and 2,500 kilograms of cargo from McMurray to Yellowknife." Wop May, now superintendent of Canadian Airways in Edmonton, decided to post an engineer at the main settlements along the route, as taking an engineer and his baggage on each flight took up valuable cargo space. Flying down north was big business, but it never got easy.

Flame Pots, Mukluks, and Caribou Stew

Keeping the planes in the air was the job of the engineer. Glory was reserved for the pilots, but the engineer was the first up in the morning and the last to bed at night. The job was hardest during the long, cold winters. Walter Gilbert, in his memoir of northern flying, recalled how the cold could turn rum almost solid: "I have seen it shaken out of a bottle almost like tomato ketchup."

Each evening, the engineer drained the oil from the plane's engine. Covers were placed over the wings, and a canvas cover hung over the engine. The latter was usually banked with snow at the ground to anchor it. Flame pots were filled with gas, ready for morning when, after a hurried and solitary breakfast, the engineer made his way to the shed near the planes and lit the wood stove to heat the engine oil.

Pouring warmed oil into the oil tank of Mackenzie Air Services' Eldorado Radium Silver Express before starting the engine on a cold winter day. The Bellanca's nose is covered by the engine heating tent. Courtesy Provincial Archives of Alberta, A12,038

Outside, he crawled under the engine cover and lit the flame pots, keeping an extinguisher nearby in case the engine caught fire. It took about an hour to thaw the engine, then the heated oil was brought out and poured into the oil tank. The engineer quickly removed the engine cover, and by then the pilot was usually there to start the engine. The wing covers came off next, and were stowed with the engine cover inside the plane. Each time the plane landed and the engine was turned off, the engine cover went on. The task was doubled or tripled on two- and three-engine planes.

Engineers, or mechanics, needed patience, a good collection of expletives, and the warmest clothes possible. Layering allowed you to adjust the amount of clothing to the temperature. "From the skin out," Russ McLaren remembered, "it went like this: light-weight pure wool underwear with long legs and long sleeves, a cotton shirt, a sleeveless wool sweater, another shirt of heavy material, melton ski pants, wool windbreaker, two pairs of heavy socks or a pair of duffle socks, felt insoles, moose-hide moccasins, parka, wind pants and assumption sash." A ski cap with ear flaps or a leather flying helmet completed the ensemble.

Wool breeches and mukluks could be worn instead of ski pants and moccasins. The assumption sash, wrapped twice around the waist and then knotted, kept the wind from whistling up under the parka. The wind pants, made of a light, finely woven material cinched at the waist and ankles, could be pulled on quickly when a wind came up. Mittens were of moose hide with duffle liners, and were worn over woollen gloves or mitts. The moose-hide mittens, using a system despised by school-children everywhere, were tied together by a rope that went over the shoulders and around the back of the neck.

Getting the planes going in the morning and putting them down for the night was only one aspect of the engineer's job. He also had to keep them in the air, a job involving hundreds of repairs depending on the flying schedule, the temperature, and the age of the planes. Broken propellers, snapped struts, ruptured fuel pumps, damaged fuselages, carburettor adjustments, damaged shock legs: virtually anything could, and did, go wrong. The fuselage could be torn by flying shards of ice or hard snow, and the mechanic had to lie on his back and stitch the tears shut with a long needle held in bare hands. Floats could be punctured by logs or rocks beneath the water's surface. Wing fabric could stretch. Winter repair jobs were hampered by the cold and the wind, while summer brought swarms of biting insects.

Engineers also had to schedule the major overhauls required by civil aviation regulations. All problems and repairs were noted in the plane's log book, which recorded the time and duration of every flight, the name of the pilot, and other details such as the amount of oil and gasoline consumed, the weight of the load carried, or the purpose of the flight. Mechanical problems that cropped up during a flight were noted, as were the remedial actions taken. The log book for Commercial Airways' Bellanca Pacemaker CF-AKI has many such entries: "controls and stabilizer stiff, repaired stabilizer; oil pressure too high, repaired at Arctic Red River; broke tail skid, had welded at halfway, broke tail skid, had to have rewelded . . . broke tail skid, had one made at Chipewyan; overflow, ice on bottom of skis, removed skis, placed two 1" hickory planks under pedestals, rebent tail skid attachment; rudder checked, control cables tightened; cleaned oil pressure line."

The engine and carburettor required the most regular adjustments and the most work. The log book of Fokker G–CAHE, for March 1933 records: "Carburettor trouble, removed carburettor and found R jet had become unscrewed. Replaced jet and checked carburettor, engine OK; oil pressure down, checked oil pressure relief valve, oil screen, engine OK; tightened carburettor, corrected R magneto points, checked tappets, oiled valve springs, checked cowlings . . . checked carburettor float and jets. Engine OK."

Major overhauls took place during the in-between seasons, and each time the aircraft was inspected for airworthiness. The overhauls were exhaustive, as an entry in one log book for the spring of 1931 indicates. Among many other procedures, the engine was taken out and sent to the factory for a complete overhaul; the wings and tail assembly were dismantled, and each part taken apart and inspected; all the nuts and bolts were tightened, and any wooden parts such as the leading edges were repaired if they required it; the plane was put on floats, and after the floats had passed inspection, they got two coats of paint; and

Repairing an airplane in the midst of a northern winter was a challenge. Courtesy Provincial Archives of Alberta, A5855

Installing pontoons on a Fokker Super-Universal. Airplanes were often lifted by a tripod and a block and tackle to change the landing gear. Courtesy Provincial Archives of Alberta, A5844

finally, the engine was put back in and the whole package tested.

If a plane was in an accident, the urgency of getting it back in the air put extra pressure on the engineer, for each hour out of the sky meant either lost business or more work for the other planes. When Canadian Airways' Junkers CF–ARI went through the ice at Fort Chipewyan in December 1935, the undercarriage and lower front fuselage were submerged. Coping with cold temperatures, thin ice, and frequent partial submersions themselves, Don Goodwin and Frank Kelly worked to salvage the plane. They set up two tripods, one at the front of the plane and one at the rear, and gradually lifted it from the water, removing damaged parts as they emerged. Salvage operations continued all day, then in the evening the damaged parts were dismantled and repaired, or rebuilt. When the temperature dropped enough to freeze the ice hard, the plane was pulled to shore by a team of horses. The repair work continued into January, with a successful test flight on the 9th. The work of the two engineers, assisted by the superintendent of maintenance, Tommy Siers, had been matchless.

Northern flying was taxing on the planes and could be perilous to the people who flew and serviced them. In the summer, the companies squeezed every aerial minute they could out of the long hours of daylight. In the winter, most planes carried emergency food supplies and other gear to increase a crew's chances of survival if they were forced down by the weather or a mechanical failure. The emergency stash might include rifles, bedding, tent, fire pots, tool kit, engine parts, axe, shovel, tarpaulin, and snowshoes. "Airplane pilots may have a lot of fun in picture shows and story books," one pilot declared, "but for real, worthwhile, he-man enter-

When one of Wop May's planes went through the ice at Fort Chipewyan, the only way to transport the salvaged parts was by horse-drawn sleigh. Traditional forms of northern transportation remained crucial after the introduction of the airplane into the transportation system. Courtesy Glenbow Archives, Calgary, NA-1258-49

tainment," they should try flying down north.

Northern flying could also be a lot of fun. The pilots and mechanics and their families were warmly received in northern communities. Jeanne Gilbert, arriving at Fort McMurray in January 1930 with her husband Walter, settled into a two-room wooden house. She ordered a table, chairs, cutlery, and pots and pans from the Eaton's catalogue, painted the walls and hung some pictures. In a letter, she wrote of spending time with Wop May's wife, Violet, the two of them huddled around Vi's radio for news of their husbands. She also wrote that the people were very nice, and had been very kind.

It was the same story for the flyers in the communities along their routes. Con Farrell and Russ McLaren spent New Year's 1938 with the Hudson's Bay Company manager at Fort Chipewyan. They shared a grand turkey dinner at the home of a Mrs Woodman, then moved on to a party where everyone reeled, jigged, and square-danced the New Year in.

Bars and hotels were homes away from home for crew members, and the flyers developed an easy camaraderie. RCMP officers, missionaries, and other residents opened their homes to the flyers when overnight accommodation was needed. Of course, some accommodations were not the best—one crew of fish-haulers had to put up in a mud shack—and the food could become monotonous as the winter wore on. McLaren recalls that some northern hotels served caribou porridge for breakfast, caribou stew for lunch, and roast caribou for dinner. It was too expensive to have fresh supplies flown in.

There were dangers, too: the image of Jeanne Gilbert and Vi May huddled by a radio is a vivid illustration of the strain of having a loved one flying down north. Ice

Search and Rescue

Flying in northern Canada meant dealing with the country's most inhospitable conditions. The danger of those flights, and the feeling among the pilots that they were doing important work, forged them into a northern fraternity. When a plane went down, other flyers felt duty-bound to search for it. The searchers, in turn, faced the same conditions that had put the plane down in the first place: vicious snow storms, difficulty in remaining on course, and poor visibility being among the worst. In one instance, six planes from four different companies searched for 105 hours and logged over 46,670 kilometres in the search for the MacAlpine party in late 1929. This party of explorers, led by Colonel C. D. H. MacAlpine, who was involved with NAME, had hoped to fly over much of the north, using Aklavik as a base. When they made it safely to Cambridge Bay, efforts then switched to flying them back to Winnipeg. When a plane went down, according to Ernie Boffa, "You don't come back until you find him."

on a plane's wings could send it down; a patch of thin ice on a frozen lake could send it to the bottom. Maps were often neither detailed nor accurate, and compasses were less and less useful the closer you got to the north pole. Engines could stall in mid-air. Fuel gauges weren't always reliable. Storms could appear from nowhere, and only a fool would try to fly through them. In the summer, "glassy water" was a hazard, as a combination of dead calm and early morning or late evening light could make it impossible to judge the plane's height relative to the water. Pilots had to throw something out the window to break the surface of the water and make ripples. Winter white-outs made landing just as dangerous.

But the flyers wouldn't have had it any other way. It was dangerous, certainly, but they liked each other's company, and they liked being busy in a job they loved during a decade when a lot of Canadians weren't working at all.

Gateways to the North

"Hams, bacon, lard, sausage and butter in sealed tins, manufactured and packed especially for northern requirements," declared the ad for Burns and Company in 1933. Burns called itself the "home of service to the northland." Other establishments might have quibbled with that. J. H. Ashdown, a western hardware store, was respected for its selection of northern mining supplies, and the Hudson's Bay Company maintained a wholesale branch in Edmonton solely for the "supply of northern requirements." Eaton's also carried a complete selection of northland supplies, including tents, guns, shells, blankets, sleeping bags, parkas, pack sacks, tump lines, tools, breeches, pants, shirts, underwear, footwear, stoves, utensils, desiccated vegetables, flour, tea, coffee, cocoa, canned meat, cameras, binoculars, first aid supplies, snowshoes, and traps.

Edmonton was once again calling itself the gateway to the north, a claim that could be supported on several fronts. It was a good place to get supplies, and it was the closest major centre to the north. It was also the temporary home of pilots waiting out the in-between times, and for miners enjoying a few days' leave. But the main arch on this gateway was the new airport, which had opened in 1930.

The 19 August 1933 edition of the *Edmonton Journal* contained a special feature on northern development. Edmonton placed an advertisement, referring to itself as "one of the important air travel centres of the world," and its airport as "one of Canada's finest." And it was just ten and a half hours by air from Great Bear Lake.

The Edmonton airport was a busy place throughout the 1930s. The aero club remained active, too, although declining revenue placed it in financial difficulty. In late 1932 the city lowered the plane storage fees it charged the club, and allowed some leeway in paying the rent that was now in arrears. The airport, however, was missing one thing, and the city knew it: a place where planes could land on water. Planes had occasionally landed on the North Saskatchewan River. They had also made use of Cooking Lake, a summer community a short distance outside the city. In early March 1933, Edmonton's engineering department proposed that a seaplane base be established at Cooking Lake. The province agreed, and bought the land. The province also built a road linking the base with the highway joining Tofield and Edmonton. The federal government made the construction of the base and landing strip part of a Depression relief program, creating work for the unemployed.

Edmonton provided room and board for the crews, and the engineering department supervised the work. Because the base would operate as part of Edmonton's Municipal Airport, the city also paid for the permanent buildings and equipment. These included an administration building and lodge, a movable dock, a slipway, a derrick for lifting the planes, and buoys in the lake for anchoring the planes once they were down.

Construction began in the fall of 1933, and by the spring of 1936 it was open. The *Edmonton Journal* covered the event in its usual manner, comparing the lodge to an up-to-date hotel featuring "smart drapes on the windows," "seven roomy occasional chairs," and "two chesterfield suites" in its lounge. A stone fireplace, magazine racks, reading and writing tables, and smoking stands completed the decor. There were seven bedrooms, each with two beds. The wharf was big enough for mooring "half a dozen big transport aeroplanes." The whole operation was overseen by a superintendent under the general direction of the manager of the main Edmonton airport.

Pilots, mechanics, and passengers could get accommodation and meals at the lodge. There were workshop facilities, and companies using the base erected their own storage sheds, with trucks and trailers for hauling materials, supplies, and equipment to and from Edmonton. Imperial Oil established facilities to handle fuel and lubrication sales.

Edmonton's hangar, still under construction in this photo, is illuminated by its night landing lights. Courtesy City of Edmonton Archives, EA–10–2770

Cooking Lake was a busy place during the summer, with vacationing families as well as the steady traffic of float-planes arriving and departing. During the in-between times, mechanics used the facilities to overhaul their planes and change the landing gear. And even during the winter, planes landed on skis if conditions at the municipal airport were not good.

Edmonton now offered complete aerial facilities. The city had entered into an agreement with Mackenzie Air Service Limited in 1935: Mackenzie leased the machine shop at the municipal airport, paying an annual rental fee and providing engineering assistance to visiting aircraft. The aero club had been providing this service, but with the increasing traffic at the airport, the arrangement was no

Children take a look at a Peace River Airways' American Eagle plane in the early 1930s. Courtesy Provincial Archives of Alberta, A2399

The Cooking Lake Seaplane Base had a number of facilities. On the right is the derrick used to lift planes to change their landing gear, while further down on the left is the Imperial Oil pump house, a dock, and a van belonging to one of the companies based at the lake. Courtesy Provincial Archives of Alberta: Alfred Blyth Collection, Bl. 242/2

longer workable. Mackenzie, which brought the airport 90 percent of its business, seemed the logical choice.

Records from the city engineering department show 193 landings at the airport and 215 at Cooking Lake in 1935. By 1936, thirty-five planes were based at the airport, increasing to forty-five the following year. The aero club was beginning to climb out of its depression as well. It was signing up new members, and older members were showing a renewed interest.

The club had always held air shows on 24 May, but for the coronation of George VI in May 1937 it pulled out all the stops. Thousands attended the civic festivities, and watched as flights of massed planes zoomed overhead. Guns boomed as a single plane dipped and looped above the cheering celebrants. The planes made several passes, and pilots likened the crowd below to "a dark smudge dotted with the whites of up-turned faces." Grant McConachie roared overhead in his Ford Trimotor; Wop May followed in a Canadian Airways Junkers monoplane. The dark green of Mackenzie Air Services also flashed across the sky, and of course the Edmonton and Northern Alberta Aero Club was there with its faithful little Moths. It was generally agreed that what the *Journal* referred to as the "great gleaming transport craft and the little buzzing Moths" had added a great deal to coronation day.

More and more planes were flying the northern Alberta skies, and they weren't all heading to the mining camps. Peace River Airways, operating out of Grande Prairie, was still active in the early 1930s, but the owner and pilot, Art Craig, finding himself on the ground more often than in the air, began to build a plane. He was also running an aviation school, and did not lack for students. Grande Prairie was anxious not to lose its airfield after the forestry patrols were cancelled, so Craig agreed to look after it. He also promised to put on a flying exhibition when the field opened for the 1932 summer season. He took people up for rides, and "Smiling" Jack Forrester, described by the *Herald* as "Grande Prairie's own parachute jumper," demonstrated his prowess, too.

Canadian Airways continued to haul mail north from Peace River by special arrangement with the Post Office. An advertisement in the *Northern Gazette* in December 1932 announced Canadian's winter air service: passengers, mail, and express would all be transported in a "fast comfortable cabin plane" that would be based in Peace River during the winter. Travellers could catch flights to Vermilion, Notikewin, Carcajou, Keg River, Hudson's Hope, Grande Prairie, Spirit River, and Great Bear Lake. Early reservations were advised.

By 1933, Peace River was beginning to think that it should be the gateway to the north, and was starting to lobby for a piece of the action. Town officials pointed out that Peace River and the east end of Great Bear Lake were "practically on the same meridian," making Peace River "closer to the great mineral fields than any other town providing railway connection with the transcontinental roads." Air traffic continued to grow in the area. After flooding cut the rail lines in 1935, food, mail, and other supplies were brought in by air. The situation reinforced the need for better air service in the minds of Grande Prairie and Peace River residents. For some time they had been lobbying unsuccessfully for better access to their communities, and the inability of all levels of government to deal with the flooding led them to conclude that they had to come up with solutions themselves. Aviation-minded local citizens, many with flying experience, decided to form a flying club. The federal government could not provide assistance during the depressed times, and an inquiry to the Edmonton club met the same response. The group pressed on, however, and in September 1935, after an inspirational speech by Major Hale, the postal inspector, the Grande Prairie District Flying Club was organized.

By early 1936, the Grande Prairie club had located a workshop and begun to build their plane. Fees were set at $25 for pilots and $5 for observers. By mid-April, one wing of the plane was on display in a local hardware store. By May, all the plane lacked was an engine, and the club undertook a fund-raising drive to buy one. The club was active at all levels. It ran training schools for prospective pilots (out-of-towners got special rates at the Grande Prairie Hotel during the course), and by April 1937 plans were in place to build a hangar at the airport. The building would house the Cirrus Moth being brought up from Edmonton for pilot training as well as the Flea the club was building. With donated lumber and volunteer labour, the hangar went up in one day in June.

Airport development in the Grande Prairie–Peace River area received a boost in the late 1930s through a general increase in aviation activity throughout the prov-

Everything for the North

Supplying northern enterprises was big business, and many companies tried to get their share. Advertisements in the *Nor'West Miner* indicate how varied the requirements were for the newly established mines and the communities growing up around them. The W. H. Clark Lumber Company of Edmonton advertised "building material of all kinds," including sashes, doors, glass, and beaver board. "We have had years of experience in packing for the North and can guarantee satisfaction," the company claimed.

"Oil Tanks for the Mines" were advertised by Karl H. Adams, a Calgary distributor. "Northerners: Why not Comfort?" queried National Home Furnishers of Edmonton, which had "a complete line of furnishings especially adapted to camp use." The McGavin bakery offered "rich, tasty, fresh" Christmas cakes and pastry made to order and delivered by air. The Great West Saddlery Company offered miners' footwear, dog harness and belting, and other leather items. Great Bear sleeping robes, "damp proof and weather proof," from The Alberta Feather and Down Company seemed a good buy, while the Edmonton City Dairy claimed that "You Strike it Rich" every time you bit into a slice of bread spread with their butter. The dairy also shipped ice cream by airmail for Christmas. From everything you needed to virtually anything you wanted, there was a business eager to supply it.

ince. In 1936, the Peace River town council agreed to develop an airfield west of town, and sought financial assistance from the oil companies and other large concerns that would benefit from it. By early 1937, United Air Transport was making regular mail runs into the district, and the *Northern Gazette* reported that more and more people were travelling by air. Fares were being reduced, and the booking agent commented that it was now necessary to make reservations in advance.

Grande Prairie was also improving its field in response to more and more aerial activity. When the first mail plane landed in March, the pilot was greeted by the mayor, the president of the Board of Trade, and the postmaster. An inspector from the Civil Aviation Department looked over the field in May 1937, and declared that there was no reason it could not become one of the best in the country. Work began at once to extend the runways, level the field, and remove the remaining brush and trees. A small house was moved onto the field as a waiting room, and a telephone was installed.

The announcement of weekly service between Edmonton and Whitehorse in April brought even more excitement. The route included stops at Grande Prairie, Fort St John and Fort Nelson in British Columbia, Lower Post on the British Columbia–Yukon border, then over the Cassiar Mountains to Whitehorse and Dawson. On the inaugural run in July 1937, over 100 people turned out to greet Grant McConachie and have a look at his Ford Trimotor as he landed to refuel at Bear Lake, just west of Grande Prairie.

In 1938 a new runway was built at the Grande Prairie airfield, bringing the total

Loading Yukon Southern's "Yukon King" in Edmonton. The nose opened and parcels were placed inside. Courtesy Provincial Archives of Alberta: Dan Campbell Collection, C158

number to four. Planes could now take off and land in any direction. Vancouver became accessible on Grant McConachie's airline for $89 return. There was so much activity in the area that a new line, Peace River Airways, emerged in the spring. The company was locally run, financed by local shareholders, and had a board of directors selected from nearby communities as well as from Peace River. Its establishment was a direct response to the belief among community leaders that the district was being unfairly bypassed by the northward trade.

The president of the new company was also the president of the Chamber of Commerce. By forming its own air company, he said at a banquet celebrating the first flights of the new air service, Peace River would show the aviation world that it was the true gateway to the northern mining country. The time had come to break the stronghold of the Edmonton Chamber of Commerce so that Peace River would come into its own as the gateway to the north. It was close, freight rates to the Peace were low, and air fares from Edmonton to Yellowknife were high. In September the company bought a new Waco, described in the *Northern Gazette* as a "luxury airliner, which provides the utmost in comfort for air travel." The plane would be available for viewing on Labour Day, at which time it would also be taking people up for rides "to demonstrate the luxury of travel by air in a modern passenger plane."

This Christmas card from Canadian Airways shows many of the company's destinations as well as its varied work. Courtesy City of Edmonton Archives, A81–26

Peace River Airways joined Yukon Southern Air Transport, Canadian, and Mackenzie in providing service to the Grande Prairie–Peace River district. The first hint of spring breakup in 1939 found Peace River still clogged with air traffic as companies tried to get in as many flights as possible while the ice was still thick enough to land on. The newspaper reported that pilots were flying from dawn until dark. During 1939, Yukon Southern brought new twin-engine Barkley-Grows into service on the route, and the fast and flashy planes travelled from Peace River to Edmonton in an hour and fifteen minutes. The first Barkley-Grow on the Edmonton–Vancouver route that stopped in Peace country was named the Yukon Queen. "Fully sound-proofed and comfortably fitted," the *Calgary Herald* declared, "it brings to residents of the Peace River district speed, comfort and security equal to that obtainable on any airline in the world." By August, Yukon Southern had three Barkley-Grows on its routes through the Peace country.

Peace River never realized its dream of replacing Edmonton–Fort McMurray as the route of choice to the north, but the increased activity did give the local economy a boost.

It was the formation of the Alberta and Northwest Chamber of Mines in 1936 which, by strengthening Edmonton's claim to be the gateway to the north, more or less discounted all other claims. The chamber was by its very existence another indication of the amount of northern flying that depended on Edmonton's facilities. It was a link between governments, mining companies, businessmen, and miners. It kept up-to-date maps, reports, and staking plans on hand for easy reference by anyone interested in northern mining. Mining equipment catalogues were also available, and the chamber maintained a register of workers, from technical experts to miners.

The chamber was instrumental in the campaign to upgrade Edmonton's facilities. The city had asked the federal government for assistance in expanding the capacity of the airport, arguing that Edmonton was bearing the burden of the development of a

"Ceiling and Visibility Unlimited"

These were the words that pilots and their families loved to hear, for it meant that flyers could expect clear conditions ahead. CAVU, as it was known in radio call letters, was also the name adopted in 1940 by a group of women whose husbands were involved in northern flying. Called the Ceiling and Visibility Unlimited Club, the group combined wartime fund-raising with providing support for each other when their husbands were off flying. The group continued its charitable and social activities following the war. Kitty Moar, the first president, explained the motives of the organization in an interview with the *Edmonton Journal* in 1990: "Many members were young wives who were desparately lonely and whose husbands were going into the most precarious flying situations. We understood one another and laughed at the same things." For everyone involved with flying down north, CAVU was music to their ears.

major part of Canada, and municipal finances should not have to shoulder the full expense. The first priority was a new hangar that would hold all the planes now based at the airport. The Chamber of Mines, together with the Edmonton Chamber of Commerce, wrote a letter to the Honourable C. D. Howe, Minister of Transport, quoting some statistics. At Edmonton's airports in 1935 alone, Canadian Airways and Mackenzie Air Service had carried over 254,000 kilograms of freight, United Air Transport 352,000 kilograms; the three companies had carried a total of 3,886 passengers; planes based at the airport were valued at over $650,000 but only six could be accommodated at one time in the hangar. "It is therefore imperative," they wrote, "that steps should be taken to accommodate these planes and house them for the safety of northern transportation." The *Edmonton Journal* also bemoaned the inadequate hangar facilities and the grass runways, which were miserable when wet, and even when dry were increasingly unable to handle the heavier and faster planes now coming into use. If Edmonton wanted to keep its role as an air centre, proper facilities had to be provided.

Activity in Calgary was also starting to pick up after a number of difficult years. Annual reports for 1933, 1934, and 1935 show little activity and a growing feeling of despair. Rumours of the resumption of airmail flights circulated from time to time, but they were fuelled more by hope than substance. Calgary continued to lobby for the air route over the Rockies to Vancouver despite persistent rumours and even hard evidence—airstrip construction by federal relief projects—that the Crowsnest Pass would be the favoured route when transcontinental flights resumed. Calgary's

Shots of early interiors are rare. This picture of the hangar at Calgary's airport in 1937 shows doors leading to the instruction room, the rigging shop, and the engine room. The middle plane, possibly under repair or being used for instruction, is missing its wings. Courtesy Glenbow Archives, Calgary, NA–4670–2

airport supervisor felt abandoned by federal authorities; years of municipal cooperation in furthering the cause of aviation and in developing facilities were being ignored.

Some seeding and levelling of the landing courses were completed by men on relief, and construction of a new runway was begun in 1934. The men used shovels and wheelbarrows, and had the occasional use of two trucks borrowed from the Electric Light Department. The work proceeded slowly. The Calgary Aero Club had two planes in operation in 1935, and gave passenger rides at the Calgary Stampede and from Chestermere Lake. In 1936 the situation began to improve as the hangar got a fresh coat of paint, the aero club got a new Tiger Moth, and the *Calgary Herald* started a column entitled "Airport Activities."

But it was Edmonton that led the campaign for expanded facilities using federal money. What neither Edmonton nor Calgary knew was that the federal government had plans of its own, which C. D. Howe and his powerful new Department of Transport would soon be putting into place. The Liberals were about to embark on a grand transcontinental transportation scheme.

From Sea to Shining Sea

When the Liberals came to power in Ottawa in 1935, they inherited the skeleton of a transcontinental air system. Under the umbrella of unemployment relief, construction of air strips had proceeded throughout the country under the administration of R. B. Bennett. Camps were established, and the men in them received room and board, clothing, toiletries, and 20¢ a day. The *Coleman Journal* reported in June 1933 that a camp would be established at the landing field three kilometres west of town to clear rocks and level the ground.

By the middle of July plans at Coleman were in full swing. The foreman and other administrators had arrived, and a cook had been appointed. Stores in the Crowsnest Pass had been given contracts to supply the camp with meat, groceries, wood, and other goods. Construction on a bath house, dining hall, and what the *Journal* described as "long accommodation tents with floors" began. Unemployed single men could register with the town clerk to be assigned to the camp.

The work of digging out the rocks, filling in the depressions, and levelling the 1,200-metre runway was slow, and came to a complete stop in the winter. Still, it was hoped that the Coleman Airport would be ready in the fall of 1935. All along the transcontinental route, projects such as the one at Coleman were being carried out. Using mainly hand tools, men chopped down trees, dug out stumps, and moved earth to build the framework of a national airway.

The newly elected Liberals decided that a new approach to the administration of transportation was needed. Various responsibilities had been spread among the departments of National Defence, Railways and Canals, and Marine. The 1936 *Department of Transport Act* brought all aspects of civil transportation under one authority. The new minister, C. D. Howe, took a particular interest in aviation. Canada had fallen behind in developing policies and facilities, and was facing a threat from American aviation concerns that wanted to expand north of the border.

The Liberals set up a subcommittee to study air policy and recommend how to consolidate its administration. That subcommittee consisted of Howe, the Postmaster General, and the Minister of National Defence. They agreed that one airline only should span the country, and once the Department of Transport had been established, Howe would entertain no other plan. Over the next few months, he studied British Imperial Airways and inspected American planes, companies, and airfields. James Richardson, head of Canadian Airways, lobbied to become the national carrier. What Howe eventually proposed was a Trans-Canada Air Lines owned jointly by the Canadian National Railways, the Canadian Pacific Railway (another of Richardson's concerns), and Canadian Airways, with these bodies and the government having seats on the Board of Directors. Richardson pulled Canadian Airways out of the deal along with the CPR, believing the government and the CNR were too closely linked for the airline to be independent. In the end, CNR took all the shares.

The *Trans-Canada Air Lines Act* was introduced into the House of Commons on 22 March 1937 and passed on 10 April. TCA began to make arrangements with municipal airfields and to build hangars and repair shops. The Department of Transport would look after emergency landing fields, lighting systems, radio communications, and meteorological services. The federal government also agreed to contribute to the upgrading of aviation facilities in municipalities on the transcontinental route.

With what would become legendary Howe bravado, the Minister of Transport announced that transcontinental service would be under way by July. This was, of course, impossible. But Howe managed to turn away criticism by staging a media event that caught the nation's imagination. At dawn on 30 July, he and two others left Montreal in a Department of Transport Lockheed bound for Vancouver. As the *Lethbridge Herald* reported, "Transport Minister Howe threw his golf clubs into a plane this morning and set off for dinner in Vancouver and 18 holes over a Pacific coast course before bed time." The plane stopped in Lethbridge briefly. Howe's flight proved that a transcontinental service was possible: its starting date was no longer an issue.

Another door was opening on aviation in Alberta, and the Depression-weary population knew it. Edmonton and Lethbridge would benefit immediately. In Edmonton, the changes occasioned by the improvements to the airport were dramatic. The size of the airfield would increase well beyond its current fifty-nine hectares, and part of a road would have to be closed. New zoning restrictions ensured a clear flight path by putting a cap on the height of buildings in the vicinity of the airport.

City council moved quickly to acquire the land and get the plans in place to apply for federal aid. The government agreed to cover one-third of the costs involved in expansion, with the exception of buildings. In Edmonton, the plans included the extension, grading, levelling, and paving of runways; fencing; installation of lights; and building taxi strips and aprons. The total cost would run around $485,500. Both federal authorities and Edmonton City Council were anxious to get the work under way.

Not everyone supported the plans. Protests about the size of the airfield and the noise from the planes had already been heard. Some residents felt that all flying activities should be located at Cooking Lake, not in the middle of the city. An alternate plan to relocate the airport to a site in Westmount Park was also rejected, as the city would lose a great deal of investment in land and improvements. The bylaw approving all the changes, and the proposed plans, was finally passed in late July, leaving Edmonton free to enter into an agreement with the federal government to begin work on the airport improvements. That done, tenders went out immediately for the contract for hard-surfaced runways.

Planes were getting bigger, heavier, and faster. In 1937 and 1938, for example, TCA acquired sleek, up-to-the-minute Lockheed Electra 10s and 14s. They needed longer runways with hard surfaces to operate safely. The runways planned in Edmonton would be 900 metres long and 46 metres wide. In September 1937, Edmonton received the go-ahead to start the improvements. The runways would be made of twenty-centimetre-thick consolidated gravel with an asphalt surface seal, which was felt to be better for Edmonton's winter conditions than a bituminous slab.

Edmonton also needed a new hangar, and tenders for Hangar No 2 were soon issued. Construction, electrical work, and plumbing all went out on contract. When officials from Trans-Canada Air Lines inspected facilities in Edmonton in early 1938, they requested that a third hangar be erected along the same lines as the one going up. TCA would then rent it from the city, and pay field usage fees as well. The company would run the hangar and pay for lighting and heating, while the city looked after maintenance costs. At the end of the lease, if TCA felt the hangar no longer suited its purpose, it would revert to the city and TCA would construct its own facility.

In Lethbridge, there was relief and enthusiasm at the announcement of the new transcontinental service. Flying in Lethbridge had not come through the Depression well, and the prospect of renewed activity was good news indeed. Lethbridge would be the last stop on the route before the planes headed over the Rockies, and the point at which flights from Edmonton, and later Calgary, connected with the national system.

A large grain terminal on the edge of the existing airfield meant that a new site had to be selected for the airport. The city purchased land south of the city from the Canadian Pacific Railway, and construction began in July 1937. Four runways were surveyed and graded, and two were prepared for hard surfacing. TCA built a hangar and office facilities, providing room for the radio and meteorological services required by the Department of Transport. Construction of the hangar began in December 1937. Local bricklayers, carpenters, and other tradesmen were used, and as much material as possible was purchased locally, giving the Lethbridge economy a welcome boost.

Various TCA and Department of Transport employees who would be stationed at Lethbridge began to arrive with their families. Pilots, air engineers, meteorologists, and radio operators adopted Lethbridge as their home. Pilot-training flights between Lethbridge and Winnipeg started that December using new Lockheed Electras. The *Lethbridge Herald* described the planes as "giants" with "power, poise, and purpose."

The new airport, named Kenyon Field after the man who had flown the first airmail into the city earlier in the decade, was a pilot's dream, the approach unobstructed from all sides. Operations began in October 1938. The official opening on 7 June 1939 was a grand affair. The *Herald* published a special aviation number to celebrate the event. A message from C. D. Howe declared Kenyon Field "one of the finest aerodrome sites on the continent." In a congratulatory message, the vice-president of TCA declared that "nowhere do I find more enthusiastic air-mindedness than in Lethbridge."

Service on the transcontinental route was phased in gradually. In March 1938, airmail service between Winnipeg and Vancouver began on a trial basis, with a stop in Lethbridge. Bad weather interrupted the initial trial flights, and it took three tries before a plane finally hopped over the mountains and swept into Vancouver. A week later there were successful flights going east and west between Winnipeg and Vancouver, with stops at Lethbridge. The service was removed from trial status and formally inaugurated in October.

In October 1938 a feeder service between Edmonton and Lethbridge was added. Pilots and ground crew began arriving in Edmonton in September, and undertaking practice flights along the route. Plans for the first official flight took on a celebratory flavour. Mackenzie Air Service would fly from Port Radium to Edmonton, picking up mail along the way. The old mail coach that had traversed the muddy trail between Calgary and Fort Macleod in the last century was taken out of storage to transfer mail from the Edmonton Post Office to the TCA plane.

The resumption of airmail service rekindled something of the wonder of flight in many Edmontonians, as well as re-establishing that feeling of community with the rest of the nation first felt with airmail service in the early 1930s. Flying down north was exciting, but more people had letters and parcels going to friends and relatives in other parts of Canada and across the Atlantic. The *Edmonton Journal* as usual caught the excitement, and added a touch of the poetic to the proceedings: "Cheered by 4,000 Edmontonians, by little ones gazing wide-eyed, by old-timers who pioneered when the resurrected mail coach that brought the mail from the Post Office was plying its lonely route from Calgary to MacLeod, a twin-engine TCA ship inaugurated the airmail run from Edmonton to Lethbridge Saturday night. . . ." The following night, the paper reported that the flights had already become routine, and there were no crowds to watch the arrivals and departures, although the reporter couldn't resist musing on "far cities drawn into a small circle by the miracle of flight."

The Calgary airport was not suitable for large planes. TCA officials had inspected the facilities, and found the runways too short and the approach obstructed by high-tension wires. As in Lethbridge, a new airport was needed. Land north of the city was purchased in August 1938, and construction began. Runways and roads were graded and gravelled, and the city borrowed money to build a hangar and administration building. When the runways were ready, Calgary was added to the Edmonton–Lethbridge feeder line on the airmail service. The first TCA flight landed on 1 February 1939, and over 1,000 chilly people watched the first plane going south with over 900 kilograms of mail.

The new hangar and administration building were ready in June 1939. The airport supervisor was proud of the new site, and noted in his annual report that many improvements, including planting shrubs and grass, had been carried out. "We are credited," he wrote, "with having the most pleasing and business like airport in

Canada." Apart from TCA, however, few planes made use of the airport.

TCA began passenger service across Canada in April 1939, and Edmonton was looking forward to establishing across Canada the kind of aerial relationship it had long held with the north. In Calgary that year, the airline brought in 3,835 passengers and 11,182 kilograms of airmail, and carried out 3,612 passengers and 12,273 kilograms of airmail.

Department of Transport regulations required a full weather forecasting service on the trans-Canada route, with pilots in contact with the weather stations and the stations in contact with each other. While this permitted the fast and accurate transmission of meteorological information, it did not prevent the weather from doing what it did best: throwing fog, sleet, wind, clouds, and snow in the paths of the planes.

Weather forecasting was one part of what the aviation world referred to as aids to air navigation. The other part was radio services. The 1938 *Canada Yearbook* describes how Canada's transcontinental aids to air navigation system complied with the standard system generally in use in North America. Every terminal airport, defined as an airport at which regular stops were made, was to have weather reporting facilities and a radio-beam station. The radio beam guided the pilot onto the airport's landing path and brought him in at the right angle of approach. Each airport's weather personnel were in contact with their counterparts across the region and with the central bureau. Forecasts were issued four times a day, supplemented by the local observer's calculations. At Kenyon Field, for example, an anemometer recorded surface wind velocity and direction while hydrogen balloons, released from the roof of the hangar, measured wind activity above ground. Information also came in over

A smart new TCA mail truck in Calgary, 1939. Courtesy Glenbow Archives, Calgary, NA-4669-7

the telephone or on the teletype. Detailed weather information was relayed both to the pilots and to the next station on their route.

The cost of radio and weather facilities was carried by the Department of Transport. The department also continued to provide funding to upgrade civil aviation facilities after TCA was up and running. In mid-1939, it financed over $45,000 worth of drainage improvements to the runways at Edmonton, as well as approximately $19,000 for various other projects.

Trans-Canada Air Lines urged the public to "Save Time! Save Money!" In an effort to attract train travellers, the airline promised "no berth costs, no meal charges, no tipping." A TCA schedule issued in November 1939 outlined the airline's routes and connecting flights, affirmed that TCA office staff would assist passengers with hotel reservations, and listed taxi information for Lethbridge, Calgary, Edmonton, and other terminal cities. Passengers were advised to wrap fountain pens before flights in case changes in atmospheric pressure should force the ink from the reservoir. Each passenger could carry eighteen kilograms of baggage without charge, but animals, birds, or reptiles could not account for any of it.

The *Transport Act* of 1938 established the Board of Transport Commissioners to license air routes, set tariffs, and administer other economic aspects of civil aviation. With the establishment of a national air service, international air travel now seemed tantalisingly close. Lethbridge hoped connections with Great Falls, Montana, would open the "Pine to Palm" sunshine route that would stretch from Alaska to the southern United States. And all of a sudden, too, visiting the Orient, or Paris, or Rome seemed possible.

More seemed possible at home as well. Flying down north was still going strong. Mackenzie Air Service was calling itself "one of Canada's leading Air Transportation Companies." The colourful green and gold Mackenzie fleet offered service to mining areas and trading posts in the Canadian northwest from Edmonton to the Arctic. Further south, membership in flying clubs was starting to rebound, and the Calgary Aero Club offered instruction leading to a private pilot's license for $175. The Provincial Institute of Technology and Art continued to offer a two-year course in aeronautics, training students in the construction, repair, and maintenance of aircraft.

In Alberta, aviation had emerged from the Depression as an established part of everyday life. People, mail, perishable food items, and mining equipment regularly went by air to northern Canada from every province in the country. Connections were made from several Canadian centres with flights heading for the United States and ocean liners steaming for Europe. Planes were bigger, faster, more comfortable, and certainly more reliable. By the late 1930s flight had become less a romantic quest than an everyday occurrence.

The progress of aviation in Alberta could also be seen on the ground. The 1938 *Canada Yearbook* contained a map of the Trans-Canada Air Lines that identified all the airports, aerodromes, and emergency fields associated with it. In Alberta, there were twelve locations: Medicine Hat, Bow Island, Taber, Lethbridge, Macleod, Coleman, Kirkcaldy, Calgary, Carstairs, Mayton, Ponoka, and Edmonton. Further north, Fort McMurray, and communities in the Grande Prairie and Peace River areas boasted aviation facilities, too. Edmonton's Cooking Lake seaplane base was merely a developed example of what many lakes in the north were: liquid airfields.

By 1939, civil aviation was a growing source of employment. Many small companies in Alberta lost employees to TCA. But another, grimmer trend was also becoming apparent in flying circles: there were more and more jobs in military aviation. Expanding air forces in Canada and Britain had taken many graduates of Alberta's aeronautics courses. Others had been hired by busy aircraft factories. Alarming events in Germany, Italian aggression in Ethiopia, the Spanish Civil War, and the failure of the League of Nations to find peaceful solutions to international problems created a tense, unstable world in which the airplane might play a devastating new role.

Hitler invaded Poland on 1 September 1939. On 10 September, Canada declared war on Germany, and flying in Alberta changed forever.

The World in Our Back Yard

Canada's declaration of war came one week after those of Britain and France. Prime Minister Mackenzie King waited that long to emphasize a point: Canada would make her own decisions as an independent nation. While King's concern with Canada's independence on the international stage was primarily directed at Great Britain, it was also directed internally. By recalling Parliament to debate a declaration of war, King hoped to convince Canadians that a concern for national unity would influence all decisions taken on matters of war. In King's mind, national unity would not be served by supplying ever-increasing numbers of soldiers to fill the places left by the wounded and the dead. In short, he wanted to avoid conscription, the issue that had split Canada during World War I. It seemed natural, therefore, that Canada should focus much of its efforts on the war in the air.

The idea of cooperative air training had first been raised at Imperial conferences in the 1920s. The Royal Air Force and the Royal Canadian Air Force had been cooperating for some time. Canadians were accepted into the RAF in limited numbers, and by 1938 British recruits in small numbers were being trained in Canada. The gathering war clouds threatening Europe by the late 1930s, led by the resurgence of the German air force, caused Britain to think about expanding its air training capacity. Far-off Canada appeared most suitable for the purpose.

During the protracted negotiations to formulate an air training plan, and in the face of British disbelief and obstinacy, King refused to abandon the two principles that would direct Canada's participation in the war: independence and national unity. A master at negotiation, he added a third: Canada would not participate merely as a contributor. If there were to be economic benefits arising from this war, Canada would get her share.

Finally, on 17 December 1939, the agreement was signed, and the British Commonwealth Air Training Plan (BCATP) was born. Canada, Great Britain, New Zealand, and Australia were the signatories. King was proud of his efforts. He had succeeded in arranging what he felt would be the major Canadian contribution to the war effort, one that avoided a commitment to supply a large land force. Furthermore, he had held fast to his demand that the plan be run by the RCAF, and that the creation of separate Canadian operational squadrons was possible in the future. In the interim, shoulder badges would indicate members of the RCAF. King had also wrung financial concessions out of the British, including an agreement to help Australia and New Zealand raise the Canadian dollars necessary to pay their way in the plan.

The plan would be run by four training commands that neatly carved the country up into zones: No 1, centred at Toronto, administered western Ontario; No 2, centred at Winnipeg, covered Manitoba, part of Saskatchewan, and part of northwestern Ontario; No 3, headquartered at Montreal, administered Quebec and the Maritimes; and No 4, its Regina headquarters moved to Calgary in September 1941, ran the program in the rest of Saskatchewan, Alberta, and British Columbia. Each command would look after its own affairs, including supply and repair depots and recruiting.

In Alberta, the BCATP dominated aviation throughout the war years. The combination of limitless sky, clear weather, and low population density ensured that the prairies would be a prime location for air training, and flying schools dotted the southern halves of the western provinces. Alberta received No 4 Training Command Headquarters in Calgary after 1941; No 3 Manning Depot in Edmonton; an Initial Training School at Edmonton; Elementary Flying Training schools at Lethbridge (later moved to High River), Edmonton, De Winton, Bowden, and Pearce; Service Flying Training schools in Calgary, Fort Macleod, Claresholm, Vulcan, Medicine Hat, and Penhold; and a Flying Instructor School at Vulcan (later moved to Pearce). There was a Wireless School in Calgary, and a Bombing and Gunnery School in Lethbridge. Air Observer schools were run from Edmonton and Pearce. An equipment depot and a repair depot were located in Calgary.

Implementing the plan was a task of staggering proportions. The RCAF had suffered through the Depression with the rest of Canada, and at the outbreak of war had only a few planes, five aerodromes, and 4,000 members. Thousands more would be needed as recruits, to run the training facilities, train the air crews, and keep the plan running smoothly.

Before the final agreement had been signed, Canada had already placed orders for aircraft, and bases in Ontario were beginning to train instructors. The selection of potential airfield sites had also begun through the cooperation of the RCAF and the Department of Transport (DOT). The DOT inspected and surveyed potential sites, and began developing them once the RCAF's Aerodrome Committee approved the locations. The committee's basic requirements included sufficient open space and the potential for using the airfield after the war. Certain types of schools had specific requirements; bombing and gunnery schools needed a particularly large area where there would be little chance of stray ammunition damaging civilian life or property; navigation schools needed to be located near different kinds of landscapes, including bodies of water, so that students could get the varied experiences and training they required.

With war declared but the plan not yet operational, eager recruits were interviewed and put on waiting lists, to be called up when the schools were ready to receive them. Meanwhile, the RCAF's relatively new Directorate of Works and Buildings struggled to turn out the specifications and blueprints for hangars, barracks, drill halls, huts, and the many additional structures that would soon be scattered across Canada. The Supply Branch was equally busy, drawing up orders for everything from aircraft to shoe laces. The Minister of National Defence, Norman Rogers, fully understood the administrative burden of such a massive undertaking, and early in April 1940 the cabinet position of Minister of National Defence for Air was created.

The BCATP accepted its first recruits on 29 April 1940 at No 1 Manning Depot in Toronto. With the fall of France in June 1940, the plan was pushed ahead on a faster schedule, and some RAF schools were relocated to Canada. The agreement was renewed in 1942, with the RAF units being integrated into the overall BCATP. The plan stayed in effect, with few alterations, until 31 March 1945.

Those first recruits followed a plan of study that changed little over the lifetime of the training scheme. Some tinkering was done with the amount of time spent in each phase and with the air observer/navigator duties, but essentially what opened in 1940 endured until 1945. The plan's mandate was to train all the members of an air crew: pilots, navigators, bomb aimers, wireless operators, air gunners, and flight engineers. Although most of the young men who signed up initially wanted to become pilots, it was the job of the instructors and of the training regimen to ensure that only those who were best suited to the task graduated as pilots.

All recruits started their air force careers at a Manning Depot. For four weeks, they learned the basics of military life. From the Manning Depot they went on to Initial Training School to study mathematics, navigation, aerodynamics, and other

Students at wireless school in Calgary practise flag signalling. Courtesy Department of National Defence, PL 1520

subjects. The weeks spent at Initial Training School (ITS) determined the path they would follow in future training. If they were considered to have the potential to become pilots, they went on to Elementary Flying Training School (EFTS). If they were selected for another stream, they went on to wireless or air observer school. After EFTS, pilot trainees graduated to Service Flying Training schools where they received instruction in advanced flying techniques. For the pilots who graduated from Service Flying there was usually a period of leave before they were sent overseas for Operational Training. Some, especially in the initial months of the war, were sent for instructor training and assigned to one of the schools. Others received postings in the Home Defence guard.

The British Commonwealth Air Training Plan was a superb example of cooperation among the major players in Canadian aviation. The flying clubs stepped forward to run the Elementary Flying Training schools. Under the *Dominion Companies Act*, the clubs became limited liability companies. They had to indicate financial stability by raising $35,000 locally as an indication of community support. The clubs provided the instructors and some administrative and technical support, and the RCAF provided the planes and equipment. Each school also received various payments, including a set amount per flying hour, and allowances for managerial, operational, and maintenance costs. Commercial companies were also part of the plan, running the air observer schools on a contract basis. There was no profit, however; payments covered costs only.

The Order of the Day

For the fresh recruit, the first moment of life in the air force set the tone for what was to come. Potential members of an air crew needed to be young males in excellent physical shape, with perfect eyesight and a good grasp of high school mathematics, physics, and English. Age limitations changed as the demand for air crews fluctuated during the war, but in the beginning they had to be at least eighteen but not over twenty-eight. Recruits were poked and prodded in a physical exam described by one as "devastatingly thorough." He recalled spending most of the day "dressing and undressing in various offices and being subjected by impersonal doctors to highly personal indignities." The potential pilot had to demonstrate an aptitude for learning on what was termed a Classification Test. Height and weight were also taken into account; recruits could not be more than 190 centimetres and 91 kilograms.

Successful recruits went on to the manning depot. No 3 Manning Depot in Edmonton, located at the old exhibition grounds, welcomed many recruits from Alberta. The four or five weeks spent there accustomed recruits to the discipline and obedience required in military life. They received their first military clothing and various components of their kit. More physical examinations and a round of vaccinations followed. Awakened early and kept busy all day with drills, classes, and inspections, they could have had no doubt they were in the military now. Any lingering doubt vanished as each recruit was photographed, assigned a number, and ordered to make a will.

From the manning depot a few weeks of what came to be called "tarmac duty" followed until a place opened up in flight training school. Tarmac duty often meant a stint as a guard or menial tasks at a base. It was a restless time, waiting for the call to an Initial Training School. When the posting came, a new exhilaration gripped the young airman as he received the coveted white flash to be worn at the front of his cap, indicating that he was an air crew trainee.

Initial Training School was a real test as a tremendous amount of material was crammed into a short amount of time. No 4 Initial Training School in Edmonton was located in the residences at the University of Alberta. Recruits were billeted two to a room, and the room had to be kept spotless. At the morning bell they hit the floor running. They washed, shaved, folded their blankets and sheets, polished their boots and buttons, and cleaned their room. Then they rushed off for breakfast, rushed back to pick up their books, and headed to the gymnasium for the 7:15 inspection and march past. At 8:00 AM classes began; they lasted till noon. An hour was allotted for lunch, then classes resumed until 5 o'clock. Squeezed into this regimen were drills and parades, sometimes through the streets of Edmonton.

Recruits studied numerous subjects, including mathematics, geometry, rapid calculation, Morse code with both signal lamps and buzzer, airmanship, navigation, and the theory of flight. They learned to recognize the silhouettes of airplanes projected onto the ceiling of the classroom. They put on respirators and entered a building filled with tear gas, only to be ordered to remove the masks and head for the

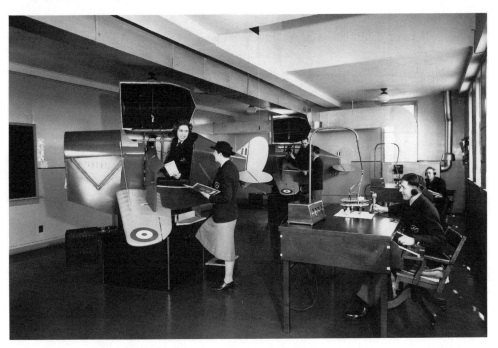

In the Link trainer, flying students learned if they had the reflexes and coordination necessary to become pilots. The instructor at the desk told the student what manoeuvres to perform. Courtesy Provincial Archives of Alberta: Alfred Blyth Collection, Bl. 520

door. From this exercise they were to learn to trust their respirators—and perhaps, by association, all their equipment. They learned about weapons and airplane engines, about meteorology and the organization of the RCAF. Room lights had to be out by 11:00 PM, but at Edmonton the lights in the shower room remained on all night. Many a worried recruit could be found there in the wee hours, hunkered down on a hard wooden bench, studying.

The stay at Initial Training School had only one real purpose: to determine if the recruit would continue on to become a pilot or move into another position such as navigator or bomb aimer. One of the most important tests came when the hopeful pilot got his first experience in the Link trainer.

The Link, a flight simulator, was the nemesis of many aspiring pilots. It looked like a baby airplane on a pedestal, but novice airmen knew that their position as pilot trainees would come to a swift end if their performance in the "crawling cockroach" was not up to par. The recruit could do almost everything in the Link that he could in an airplane. In response to the pilot controls, a series of bellows caused the Link to turn and shudder like a plane in flight. An instructor at a nearby table offered verbal directions, and kept an eye on the "crab," a machine that recorded the movements of the Link on a graph. The student could examine the graph with his instructor after each "flight" and determine where he was having problems. When the hood was pulled up over the cockpit, the student had to rely solely on instruments.

Examinations and a personal interview with the selection board completed one's stay at Initial Training School. Each man was issued two flying suits, goggles, fur-lined boots, and a helmet, and anxiously waited for the next postings to be announced. Graduation from Initial Training School, one airman recalled, "was no small thing. It meant promotion from Aircraftsman, Second Class . . . to Leading Aircraftsman. The new rank was indicated by a propeller badge sewn on each uniform sleeve between elbow and shoulder. It also meant an increase in basic pay from $1.30 to $1.50 per day plus Special Group [flying] pay of 75 cents per day, whether one flew or not." When the postings were announced, shouted in alphabetical order by an officer to the nervous men assembled before him, the joy of those headed for pilot school contrasted with the disappointment of those named to observer training. It was a difficult time, especially when good friends went in separate directions.

The recruits who had been chosen to continue pilot training went to an Elementary Flying Training School. For many in Alberta, that meant High River. Flying instruction took place early in the morning or in the late afternoon and evening. Up at 3:30 AM, the pilots-to-be headed to the locker room to await the call from their instructor. When it came, they put on their parachute packs and walked out to the plane. The instructor sat in the front cockpit, watching his student in a mirror mounted on a strut. He performed the manoeuvres he wished the student to master, and explained through a connecting tube what he was doing. The student then tried to imitate him, again listening to the instructor through the tube.

Each student kept a log book, and carefully recorded his progression from becoming familiar with cockpit layout, to taxiing and "straight and level flight." Climbing, gliding, and stalling were added, and various kinds of turns. By the fourth lesson the student was practising taking off into the wind, approaches, and landings. Finally, the student was introduced to spinning and various other tricky and evasive manoeuvres. After ten hours of instruction, the trainee was expected to try a solo flight, demonstrating at least a wobbly proficiency at flying a simple circuit.

Many trainees didn't make it as far as the solo flight, as their instructors felt they lacked the reflexes or some other requirement for becoming a good pilot. Others

The pride of a young pilot with a solo flight or two under his belt is seen in Stan Reynold's face in this photo taken during his time at Elementary Flying Training School. Stan is wearing the inner flying suit, part of the two-part winter gear. You can also clearly see the gosport, the tube with which the instructor in the front cockpit communicated with the student in the rear cockpit. The gosport plugged into the instrument panel.

Tiger Moth aircraft were used extensively in the BCATP. This photograph shows the quilted pad just beneath and to the front of the front cockpit that helped keep the oil warm. The instrument panel (with the exception of the instruments themselves) was covered with sponge rubber padding. The padding was supposed to protect the pilot's head should it hit the panel during a crash landing. Courtesy Stanley G. Reynolds, Wetaskiwin, Alberta

who weren't ready to solo at the prescribed time went for a ride in what the students called the "washing machine." This flight was taken with the Chief Flying Instructor, and was usually followed by the trainee being removed from the pilot stream—or washed out, as they called it. The curriculum was packed, the pressure to perform was high, and the time to do it was short. Awaiting the instructor's nod to go solo was nerve-wracking, but the thrill of the first flight made up for it. At High River, one recruit recalls, students who had soloed successfully rushed out to buy a polka-dot silk scarf, and could soon be seen "swanking about" in them.

More instruction followed, with more manoeuvres added, including forced landings, acrobatics, and side slipping. Elementary Flying Training School also meant more hours of ground school. Particular attention was paid to navigation and armaments, and the young pilot was tested at regular intervals. After twenty hours of solo flying, High River pilots earned the right to wear Tone-Ray aviation-style sun glasses. "When someone approached wearing a silk scarf and sun glasses," a recruit remembers, "you recognized that you were in the presence of a cosmopolitan AVIATOR, jaded with aerial adventure."

New elements continued to be added to the flying repertoire, including instrument and night flying, and cross-country flights. The names of towns emblazoned on grain elevators helped many a lost pilot find his way back to the school. Coronation was particularly thoughtful, having painted the town's name on the roof of its curling rink.

The completion of Elementary Flying Training School found those students still in the pilot stream posted to a Service Flying Training School. Here they met bigger and far more sophisticated planes such as Cessna Cranes, Avro Ansons, Harvards, and Oxfords. This was clearly advanced training. The flight course focused on formation and night flying, bombing, cross-country and instrument flying. Soloing at night was one milestone. After gaining proficiency at handling the bigger and more powerful aircraft and being declared a capable and reliable pilot, the student received permission to carry other pupils as passengers. Ground school lectures continued in navigation, armaments, and instrument flying.

A tough series of tests met the recruit at the end of Service Flying Training School: instrument flying tests, navigation tests, and the dreaded Wings Test where the pilot had to perform virtually everything he had learned. There was another agonizing wait until the list of successful students was posted. Although most yearned to go overseas, pilot graduates were assigned on the basis of need. Some stayed at home on coastal defence. Others went on for instructor training and were then posted to BCATP schools to keep the line of pilots heading for Europe and the other theatres of war unbroken. Pilots could also be assigned to pilot duties at the schools, such as towing targets for gunnery practice.

Other members of an air crew were also instructed in the BCATP. One extremely important position was that of observer. Observers plotted the exact position of the aircraft and its course, a skill called dead reckoning. Air observers spent time in ground school concentrating on subjects critical to their position: navigation, map and chart reading, direction finding with compass and wireless telegraph, meteorology, aircraft recognition, and photography. In the air, they practised it all. Then they went on to study bombing and gunnery. Successful completion of that course meant graduation and the right to wear the white O with a single wing. Four weeks of training in astro navigation was the observer's final stop before receiving his posting. Most went overseas, but some remained behind with home defence squadrons, or as instructors.

John Chalmers, a graduate of No 2 Air Observer School, explained how complicated the air navigator's job was, and how it got the designation the "gen trade." "Gen" was slang for reliable information. Where rumours and scuttlebutt were the norm, gen was something you could trust. The navigator's job was a highly skilled and demanding occupation. Giving the exact position of the aircraft to the pilot required absolute accuracy. Hence the term "gen trade." Many young men who were initially disappointed not to have made the pilot stream soon discovered the prestige a good navigator commanded.

In 1942, observer training was changed to include navigation only. A new member of the bomber crew, the bomb aimer, would now be responsible for locating the target and releasing the bombs. Navigator-bomber specialists, who worked on medium-sized bombers, and navigator-wireless specialists who navigated on fighters

Aircraft repair and maintenance were critical to the success of the BCATP. Each school had its own repair and maintenance section that kept the planes in the air. Courtesy Provincial Archives of Alberta: Alfred Blyth Collection, Bl. 529/3

and fast, light bombers such as the de Havilland Mosquito, were two other new categories. Many of the schools, including No 2 in Edmonton, expanded to meet the training load for more and better navigators as the Allied bombing campaign was stepped up.

Bombing and gunnery schools also expanded to turn out the bomb aimers and air gunners the new campaigns needed. Gunners took ground training that included machine gun practice out on a range, and classroom instruction in weaponry. They graduated to air training, where they tried to hit a six-metre canvas sleeve called a drogue, towed by a plane known as a drogue ship or target tug. The drogue was released, then picked up and examined so the gunner and his instructor could determine his accuracy.

Bomb aimers started their ground training on a "bombing teacher" that simulated the operation, then took to the air to practise dropping bombs on targets. The practice bombs released a puff of white smoke on impact, making their location easy to pinpoint.

When No 8 Bombing and Gunnery School, located in Lethbridge, opened in November 1941, it had over fifty planes in operation. Most were Fairey Battles and Bristol Bolingbrokes. This school used part of the neighbouring Blood Indian reserve as a bombing and gunnery range, and enjoyed a good relationship with the reserve. Several chiefs toured the station in 1942; they were flown over the ranges and their reserve, and entertained to dinner in the Officers' Mess. In what was recorded in the daily diary as the "surprise" of the visit, "Group Captain Jones was honoured by the visiting chiefs with feathered headdress and full regalia, and in a befitting ceremony they named this School's Commanding Officer to be known as 'Chief Heavy Shields'."

Support functions were critical to keeping the plan running. The central supply depot filled the material needs of the stations, while repair depots salvaged, repaired, and refitted damaged aircraft. Repair Depot No 10 in Calgary employed 3,000 people at the peak of the war. Six single and four double hangars housed the facilities, where overhaul and repair sections for everything from engines to radios to airframes were kept busy. Teams from repair depots retrieved damaged aircraft, and also visited stations to repair them on site. Any time an airplane went down, it was hauled back to the station, dismantled, and any undamaged parts put back into circulation. With the manufacturing capacity of the Allies stretched to the limit, nothing was wasted. Salvage, repair, and recycling were essential functions.

The motor transport section buzzed around the station picking up refuse, delivering equipment, or ploughing roads in the winter. Also included in their responsibilities were the trucks that refuelled the planes. Many stations also had tractors to tow the planes or maintain unpaved roads. Ambulances were also available, and during flying training, one was always on stand-by. Maintenance and repairs to all vehicles were done in the station's workshops.

A full staff of technicians and mechanics kept the planes in good repair. Each aircraft was inspected before its first flight of the day, and received a more thorough inspection after every forty hours of flying. One room at the station was devoted to packing and inspecting parachutes. Another, with silhouettes of planes painted on the walls, was where students learned to recognize planes in flight. Another room was filled with the log books in which every move of every plane on the station was recorded.

Changing the Landscape

On 20 November 1939, the *Lethbridge Herald* reported that five American planes had been flown from their factory to Sweetgrass, Montana. After bringing them to within a few metres of the border, the American pilots clambered out and the planes were hitched to pick-up trucks and towed across the border by Canadians. Once the planes were on Canadian soil, RCAF pilots took off from the open prairie and flew them to Kenyon Field in Lethbridge. A week later, ten more planes were delivered in the same manner to a field west of Coutts. The field was guarded by troops from Lethbridge. Interested spectators had gathered as early as 7:30 AM to watch the delivery. The planes were destined for the Royal Air Force, which was, like Canada, at war by this time. Because the United States was not at war, the unorthodox scheme of delivery was necessary to comply with the letter of international law: planes could not be flown from a neutral country to one that was at war.

More planes were towed across the border in December, and the *Lethbridge Herald* declared "Bombing Planes Worth Millions Will Clear at Border Airfields." The announcement caused a bit of excitement on both sides of the border. Aircraft of this type had not been seen in the area before. On Saturday morning, 16 December, two large Douglas bombers arrived, were taken through the now-normal procedure, and immediately took off for Lethbridge.

This small number of planes was soon dwarfed by the announcement of the British Commonwealth Air Training Plan. Communities in Alberta understood immediately what the agreement could mean. The Depression had been long and hard, and the plan offered the potential for employment and development. On 21 December, just days after the plan had been signed, the *Lethbridge Herald* asked, "Will Lethbridge be Made Training Centre for Pilots of Empire?" In the ensuing article, the paper expressed its fear that Lethbridge might be passed over. It was reported that there were three survey parties in southern Alberta collecting data in the Medicine Hat area, the Macleod district, and around Calgary. The Edmonton airport was also being surveyed, with the assistance of the provincial public works department.

Both surveying and lobbying were going on in Edmonton. The Edmonton and Northern Alberta Aero Club had been informed in September that it would soon be training pilots, and the club instructor had already left for training in Ontario. By the middle of October, five RCAF members were receiving instruction in Edmonton. All this was not lost on one of Edmonton's biggest boosters, the Chamber of Commerce's Aviation Committee. The chamber had always been a tireless supporter of aviation in Edmonton, and lost no time trying to gain Ottawa's ear. A report by G. M. Croil, Air Vice Marshall and Chief of the Air Staff, listing the advantages and disadvantages of Edmonton and Calgary as air training centres, had come to the city's attention. In the committee's opinion, the Air Vice Marshall had not been as critical in his evaluation of Calgary as he should have been, and set out to redress the balance. Edmonton, it believed, was clearly in the superior position.

The city's more stable climate, lower velocity winds, and generally better cloud ceiling made for far better flying conditions. The topography in Edmonton's immediate area was also clearly better than Calgary's. Level ground for emergency landing fields abounded near the city, while it was scarce near Calgary. The proximity of the mountains and the lack of trees made for additional difficulties for beginning pilots in Calgary, the topography in general having contributed to "several bad crashes and fatalities" according to the Edmonton group. In every aspect, apparently, Edmonton was superior. Edmonton's airfield had sewer and gas hook-ups, better rail-

Jack Manson poses proudly in his air cadet's uniform in Edmonton. Except for the high collar, it was the same as the RCAF uniform. Created by an Order-in-Council during the war, the air cadets introduced young boys to air force life. They studied aeronautics, Morse code, and other subjects necessary for a career in aviation. Many went on to join the RCAF, and found their training put them ahead of those who did not have a background in the cadets. The establishment of the BCATP created a great deal of interest in flying, and helped the cadet movement. Cadets sometimes spent time on BCATP stations, and often took part in sports days and other events. The 187th Air Cadet Squadron in High River, for example, spent two weeks at No 8 Bombing and Gunnery School in Lethbridge in the summer of 1943, and two weeks at the Service Flying Training School at Fort Macleod the next summer. Courtesy Jack Manson

way connections to the airfield, and it was not plagued by the dust storms that blew through Calgary. Heads were shaking in Edmonton: how could Croil have missed all that?

Lobbying, whether it was for air training or munitions factories, was a part of war. Edmonton, like other communities, argued for a proper share, and continued to lobby federal officials well into 1940. Writing to the Minister of National Defence for Air in September 1940, the Chamber of Commerce respectfully asked if it could bring to his attention "the experience of many pilots of wide experience who can express an opinion on the merits of locality without local prejudice." The letter went on to list several points noted by commercial pilots supporting Edmonton's superiority as a location for training facilities: nearby land was flat and free from boulders making for safe emergency landings; there were few gullies that could influence air conditions; light snowfall, no chinooks, no slush. Newspapers took up the chorus as communities all over vied to get their slice of the wartime pie.

When a decision had been made to locate a training facility in a community, it had an immediate impact. Existing structures could sometimes be used in larger centres, and some training schools were initially located in rented facilities. In Edmonton, it was at the university, in Regina, the Regina College and the Normal School. Locations with functioning airports such as Lethbridge, and those with largely empty facilities such as High River, had something of a head start. In many communities, though, everything had to be started from scratch.

That was certainly the case at Claresholm. Citizens were given a glimpse of the future when the local paper reprinted the text of a speech in the House of Commons outlining what would be happening there: hangars and barracks; mess quarters for officers, men, and civilians; canteens and recreation buildings; garages; motor and aviation fuel storage tanks; a drill hall and a hospital; storage and maintenance facilities; and a machine gun range. Department of Transport officials began in October 1940 by taking up options on land. Suddenly, quiet acreages and fields were buzzing with surveyors and construction officials. Engineers and their assistants were looking for accommodation in the town. By the end of the month, two train-carloads of grading equipment had rolled in from General Construction of Vancouver. Sixteen more would eventually arrive. Gravel crushers began filling the air with noise and dust. The *Claresholm Local Press* reported that General Construction would have sixty trucks working there by early November.

The tremendous amount of construction triggered by the plan energized the construction industry. Dirt moving, grading, runway and building construction all required a huge work force. Small communities hadn't the resources to undertake such projects, so companies from larger cities moved in, hiring local labour or subcontractors where they could. Bennett and White of Calgary, for example, had received the contract to erect the buildings at the Claresholm training station. Many of the trucks used on the project, the Claresholm paper noted with approval, were being operated by local owners.

Construction of No 8 Bombing and Gunnery School at Lethbridge is a good example of how major contractors worked with subcontractors to handle such enormous jobs. Shoquist Construction out of Saskatoon acted as general contractor for the buildings. In the ad they placed in the *Lethbridge Herald* on the opening of the school, the company thanked the sub contractors, supply houses, and employees who had made construction possible. Palex Painters of Calgary painted the buildings. Lighting and power were installed by Roy Electric. Doncaster Construction, with its head office in Edmonton, wished good luck to the boys "who will take off from the runways we were glad to construct." Freel of Lethbridge did the sheet metal work.

Stations needed other services, too. Railway spur lines often had to be extended; gas and water lines had to be laid; power, sewage disposal systems, and gasoline storage facilities had to be installed. In Claresholm, the newspaper reported that the power requirements of the station would be about four times that of the town itself, noting that a long line of transmitters would have to be erected. Construction went on continuously, as did delivery of equipment and supplies. In early January 1941, the *Local Press* reported that work was speeding up at the airport: "The gravel dump is already a small mountain. . . . High powered dump trucks rumble in and out of town on change of shifts at all hours of the day and night. Bennett and White will have several hundred men on the hangar jobs right away. The lumber and cement is rapidly piling up on the grounds, being moved by truck from the local freight yards."

Hundreds of men worked at putting up the buildings. The *Red Deer Advocate* noted that 400 men were working at the Penhold station. In Claresholm, more than 300 carpenters were needed. Bennett and White were advertising in Calgary, and hundreds of other tradesmen, from tinsmiths to plumbers to electricians, were on the job.

Each training base required auxiliary landing fields. This was especially important given the number of pilot trainees who would soon be cruising overhead. That meant more land had to be acquired; the field would have to be linked to the station

Barracks like these at Kenyon Field went up at every BCATP school. Courtesy City of Lethbridge Archives and Records Management, P19901053003

by a surfaced highway, and services such as power would have to be installed. A report in the *Claresholm Local Press* indicated that there might be some resentment in the community over the auxiliary landing field. The field blocked the Barons road, readers were informed, and would be located on "some of Alberta's choicest wheat acreage." Other areas of conflict soon became apparent. Again in Claresholm, the paper noted that some of the construction workers had neglected to pay their boarding-house bills before leaving town.

The adjustment for small towns must have been significant. Although the stations were usually located a few kilometres away, the noise and activity couldn't help but spill into the community. The promise of work also attracted legions of unemployed men. A local paper reported in January 1941 that "unskilled labour has poured in from hundreds of miles around, broke, and with no connections. The demand for these men is limited." At High River, whose air base had lain inactive for years, the contrast between the quietness of the Depression and the frantic activity must have been particularly acute.

Much of the work, although intense, was of short duration. Waves of electricians and plumbers replaced carpenters as the buildings took shape and interior finishing began. The construction process was a well-orchestrated mix of manual and mechanized labour. At Penhold, for example, crews were shingling the huts as trenching machines dug ditches for sewer and water lines. Generally, the townspeople were interested in what was going on. As the hangars were going up at Penhold, the *Red Deer Advocate* noted that, from the spectator's point of view, the "best show is the raising of the big trusses." There were twenty-two in each hangar, raised with drag lines.

As the transformation from prairie to air base proceeded, guards were posted to secure the grounds. Residents of Claresholm were warned that these boys were armed and took their duties seriously. When one guard's stern challenge to a suspected intruder went unanswered one evening, the newspaper reported, the young guard went to investigate. Supported by an officer on duty who also issued a challenge, the two shone their flashlights into the stubble before firing, thus sparing a pig and her litter from harm. In Edmonton, hundreds of residents had been watching the construction, and now had to be warned about taking pictures. New defence regulations prohibited photographing military installations.

Many BCATP stations, especially when the plan was accelerated early on, received their first trainees before they were ready. In some cases, services had not been fully installed, while in others roads and runways were awaiting completion. In Lethbridge, for example, students made do with empty nail kegs and fruit boxes for tables and chairs, and had to use washroom facilities at the Trans-Canada Air Lines hangar well over a block from their barracks. At High River, the pupils arrived only to find that "there were no roads, the ditches were open, and the whole station was a sea of mud. The hangar was far from complete and only two of the barracks had water turned on. . . . there was no gas or water turned on into the kitchen, so everyone ate picnic lunches until these services were complete." Others, like Kenyon Field in Lethbridge, underwent continuous change as the war progressed.

Lethbridge started out as No 5 Elementary Flying Training School, and several new buildings were erected to join the ones already there. In June 1941, the school was moved to High River because of the gusty winds over Lethbridge: novice pilots in their Tiger Moths found them just too turbulent to deal with. Later that year, No 8 Bombing and Gunnery School opened at Kenyon Field, necessitating an extensive period of construction to build a parade square and improve the runways for the bigger, heavier planes. More huts, hangars, and a hospital were also built. Winds also plagued the construction. Entries in the daily diary of the school read, "Starting around 0900 hours a 76 mile [122-kilometre] gale lashed the Station until well on to midnight," and "Blowing with undiminished force from early morning to 1900 hours a 61 mile [98 kilometre] per hour wind raised a terrific dust storm, making working conditions very difficult."

There was more to getting a school ready than construction. Supplies and equipment arrived constantly, and were often put in temporary storage until construction was finished. Furniture, stationary, parachutes, refrigerators, fire-fighting equipment, and aircraft poured into each station. Some of the planes were flown in, others arrived in crates by rail and had to be trucked to storage, then removed and assembled. They were all taken on test flights before going into the hangars to await their first pupils. Inspections of everything from electrical wiring to the procedure for handling stores occurred regularly. The schools also had to come to terms with local inhabitants when construction infringed on their property. The daily diary from Lethbridge shows compensation paid to a local landowner for $40 crop damage and $25 a year rental for a strip of land required for the sewer line.

Equipment and other supplies were often still arriving after training had begun. At Penhold, the *Red Deer Advocate* reported that the flying school had started "active work" even though only about twenty of the twin-engined Oxford training planes had arrived, and "most of the maintenance equipment and tools are still to come."

Opening Day, Closing Day

Opening day was the first real chance the station had to begin to integrate its existence with the life of the community. It was also the first chance for local residents, who had been denied access to the site as construction progressed, to get their first real look at the buildings and what went on inside them. Often occurring after the school had been up and running for some time, opening day was usually announced in the press, sometimes with a specific invitation from the commanding officer of the station. Civic officials sometimes declared half-day holidays to mark the event, and to make sure everyone had the opportunity to attend the ceremonies.

Vulcan's official opening was held on 30 October 1942. Anyone who subscribed to the *Vulcan Advocate* knew roughly what to expect, and the more than 4,000 people who attended were not disappointed. The band from No 2 Wireless School in Calgary played marches and Vulcan's mayor offered a western welcome to the airmen. The Lieutenant Governor of Alberta officially opened the school, and the airmen put on a show for their new neighbours, demonstrating formation flying and acrobatics for an hour while their manoeuvres were described over the public address system. The station's fire brigade and first-aid crew demonstrated their skill as well, rushing to a mock conflagration and assisting a person overcome by smoke. A "splendid lunch" was enjoyed by all.

Some of the buildings were then open for inspection. The station, comprising seven large hangars and forty-five other buildings, was compared to a town of about 2,000. The paving at the station, according to the *Advocate*, was enough for a road six metres wide stretching from Lethbridge to Macleod. The station had its own fire department, hospital, and dental facilities. 1,500 people attended the dance that finished off the day.

In High River, about 2,500 people came to see the revitalized airport. They were treated to speeches, formation flying, and aerial spins, loops, and wing-overs. Roy Lomheim, well known in southern Alberta, performed a parachute jump. Later that night the Royal Rhythm Orchestra kept the Elks Hall jumping, and a midnight supper in a hall across the street provided further fuel for the celebration.

Many opening days were held in conjunction with the graduation ceremonies of the first or second class. At No 15 Service Flying Training School, the Lieutenant Governor arrived by plane from Edmonton, and joined air force commanders from No 4 Training Command and local civic officials on the platform. Hundreds of uniformed RCAF men ringed the square. The graduating class stood at attention in the centre, facing the platform. During the ceremony, each man's name was called out, and his wings pinned on. Afterward, all ranks took part in the impressive march past led by the Air Force Band from Macleod and the graduating class. The parade had just ended as a thunderstorm sent everyone scurrying for the shelter of one of the hangars. But then, the tour of the facilities was about to start anyway!

Rollers compacted the snow on runways. This tractor is trying to deal with the aftermath of a snow storm in Penhold. Courtesy Department of National Defence, PMR 84–978

Not all the buildings, of course, were open for public inspection. Visitors were shown one or two hangars, the drill hall, the airmen's mess and barracks, and the hospital. The stores and repair sections were what most civilians found intriguing. Motors and propellers they could understand, but somehow they didn't expect to find mouse traps, spatulas, shirts, and floor polish at a military installation.

Celebrations were often repeated on the anniversary date of the opening of the station. In many cases, they were more elaborate and less formal than the opening ceremonies. The second anniversary of the High River station featured sports competitions, midway attractions, flying demonstrations, precision drills, and tours of the school. About 2,000 people attended the event on a hot fall day. They toured the classrooms, mess hall, recreation building, the barracks, and the parachute section. Visitors tried their luck at bingo, crown and anchor, and darts. There were track and field events for the airmen, and dashes for the children. Aerial acrobatics and a parachute jump completed the entertainment.

Tours were occasionally held at other times. During the national campaign to promote interest in the Women's Division of the RCAF, for example, 250 women toured the station at Vulcan. Women had been eligible to enlist in the air force since the creation of the Canadian Women's Auxiliary Air Force in 1941. In February 1942 the auxiliary was changed to the RCAF Women's Division. Commonly called the WDs, they repaired aircraft, packed parachutes, drove trucks, and performed many other duties on BCATP stations.

Carnival days put on by the stations were another way of promoting amicable relations with local communities. The "Mammoth Free Carnival" held by No 19 SFTS at Vulcan in September 1944 promised sports, a dance and floor show, games, re-

A wings parade in Claresholm in 1941. The graduating class marches smartly past the crowd that has gathered to watch. Courtesy Department of National Defence, PMR 74–280

freshments, fireworks, and raffles. The public was urged to "Come at 1:00 PM and stay until 2:00 AM."

Relations between a school and nearby communities extended far beyond these major events. Sporting events, concerts, and dances were the most common meeting grounds, as they invariably drew good crowds. Sports played a major role in the life of a base. Airmen and airwomen played inter-squad games at their own station, they played teams from other schools, and they played teams from local communities. Softball, fastball, basketball, hockey, track and field, cricket, and soccer were the most common sports, and the games seem to have been hard-fought contests with more than a little pride at stake. When the locals defeated the air force hockey team at Claresholm early in 1942, they were reported to be "a bit chesty" over the 6-2 defeat they had handed the flyers.

The *Penhold Log* for August 1943 gives an indication of how seriously the schools took their sports. The station played three levels of soccer. The Penhold Flyers were the top team, and a contender for the cup in the Alberta league. The school paper noted that the loss of the team's clever left-winger to another posting would be a real blow to their chances. The Tigers were the station's "B" team and played in the same league. The school also had its own soccer league with teams determined by section. The officers' team was currently occupying bottom place with no wins, twenty-four goals against, and only seven goals scored. Maintenance "A" was leading the league

Two student navigators head for their Avro Ansons in Edmonton, 1943. Students nicknamed these planes "flying greenhouses" because of the expanse of windows running down each side of the fuselage. Courtesy Provincial Archives of Alberta: Alfred Blyth Collection, Bl. 605/1

with twelve points from five wins and two draws.

At Vulcan, the men's and women's basketball teams from No 19 Service Flying Training School played teams from other schools and the surrounding communities. The local newspaper reported the games in detail. The airmen and women also participated in the sports days and fairs that had been held for years. At one High River sports day, the men from No 5 Elementary Flying Training School put on a precision drill display, then humiliated the locals in a tug of war by "pulling the town over the line in about five seconds flat," according to the *High River Times*.

Hockey was very popular. The airmen at High River engaged the locals in a hockey game soon after they arrived. The airmen won in a contest which, according to the *Times*, featured "fast, keen play" and "plenty of action" where "rough moments were not unusual." The air force team from No 15 Service Flying Training School near Claresholm, challenging for the southern Alberta service league title in 1942, chartered a train and took 350 supporters with them to Lethbridge for a game against the team from Macleod No 7 Service Flying Training School. The Macleod team took the series in front of what the Claresholm paper described as 2,000 wildly cheering fans split about evenly between airmen and civilians. The victorious team, now the proud winners of the Lethbridge Hotel trophy, moved on to the provincial service playoffs. The Claresholm newspaper credited the team from No 15 Service Flying Training School with rekindling local interest in hockey.

Even those who had not played the game before took part in exhibition matches to raise money for various causes. Members of the Women's Division from Macleod and Claresholm gamely put on skates and picked up sticks to help raise funds to furnish the recreation hall. One player seemed to have played the game before. Her efforts were described in the paper as being of National Hockey League style, and from "the standpoint of smooth stick handling, fast skating and excellent marksmanship, Crawford gave the fans the best hockey of the night . . . male or female." Some Australian airmen also strapped on skates to try an unfamiliar sport. Their biggest enemy, according to the published account, was gravity: there "wasn't a pair of hockey shorts that didn't have snow on them when the game was over."

Boxing was particularly popular as a spectator sport with civilians. Boxing cards often featured enlisted men as competitors and drew tremendous crowds and extensive press coverage. Residents of Vulcan and district were invited to a mammoth boxing card at the airport in January 1943, announced on the front page of the newspaper: "Headlining the biggest sports attraction staged in Vulcan in a long time, Pte. Al Lust, the sensational Western Canadian contender for the Dominion welter title takes on Flt. Sgt. Billy Evans, RCAF in the main bout of a star spangled boxing and wrestling show." An additional twelve bouts were also featured, and the "grunt and groan fans" would find satisfaction in the wrestling contests. Men paid 50¢ admission; women got in free.

Servicemen were constantly travelling to other stations to box. An evening of boxing drew 1,200 fans to the drill hall at No 2 Flying Instructor School near Pearce in February 1943. The all-service card featured boxing, wrestling, and a fencing exhibition put on by a member of the Women's Division from Lethbridge and an airman from the RAF.

Most stations also enjoyed sports in less organized forms. There were tennis courts, volleyballs and nets, baseball gloves and other equipment for idling away the leisure hours. In High River, the airmen fixed up the local swimming pool, to the delight of the residents. High and low diving boards, a covered shelter for picnics, and a paddling pool for the children made it a favourite destination for people both from the base and from the town.

Concerts were another important form of organized entertainment, and promoted interaction between the school and the neighbouring communities. Most stations had bands that played at station functions, concerts, and in local parades. They could also be found whipping up crowd support at sporting events. Some stations had orchestras as well.

Local dance and music students from the civilian communities often put on shows for the airmen and airwomen, who were often no strangers to the spotlight themselves. Skits and comedy sketches, often based on the tribulations of air force life, and sometimes full-length musicals were part of the repertoire of many an air force troupe. They were an effective means of boosting morale, and the air force put together shows from the best talent in the force and toured them to all its schools and bases.

The small towns and the air force also cooperated in providing recreation facilities

The band from No 7 Service Flying Training School at Fort Macleod awaits its turn on a radio broadcast over CFAC. Bands were one of the many activities that kept trainees and staff members busy, and fostered good relations with local communities. Courtesy City of Lethbridge Archives and Records Management, P19911022007

for those on the stations. The BCATP provided for recreation halls on the stations, but the furnishing and running of them were left to local authorities. Organizations as varied as the Legion, the YWCA, the United Farm Women, the Eastern Star Lodge, the Sunshine Club, and men's clubs and service organizations took part in organizing and running the facilities. They raised money, sewed curtains, collected donations of sports equipment, books, and magazines, and dispensed soft drinks, candy, and sandwiches from the canteens.

In Lethbridge, the War Services Auxiliary worked with the Legion and other local groups to furnish the recreational rooms at the school. The reading room and library featured works of fiction, magazines, technical books, newspapers, and an encyclopaedia. Stamps, money orders, and a telegraph service were available, and the Canadian Legion War Services provided writing paper and envelopes. As described in the *Lethbridge Herald*, the room was furnished with "bright drapes, interesting pictures, plants, chesterfields, occasional chairs, rugs, writing desks and tables, ash trays and reading lamps." The room was meant to be a home away from home for service men and women. The recreation hall at the station near Claresholm featured a fireplace, a piano, a juke box, and a radio, and was well stocked with cards and board games.

The YWCA opened Hostess houses at bases and nearby towns across the country. In Alberta, the first one opened at No 15 SFTS in Claresholm in June 1942. The Hostess House provided some of the same functions as recreation and reading rooms, including writing materials, books and magazines, games, and a canteen. The YWCA provided more personalized service as well, visiting the sick and the lonely, helping with wedding arrangements, and assisting the newly arrived wives of airmen.

Meeting places were often developed in the towns by local people. The Recreational Room Committee in Claresholm canvassed the district for funds to extend the IOOF Hall to provide a meeting place for airmen and airwomen. As the fund-raising proceeded slowly, construction was undertaken in stages. Local women's groups volunteered to take turns keeping the hall open and running the canteen. The hall

Visions of Sugarplums . . .

Christmas far from home could be difficult for young BCATP recruits, and bases usually did all they could to put on a good holiday meal. At Penhold in 1942, the Christmas dinner was quite an affair. And as was customary, the officers served dinner to the lower ranks in the airmen's mess. The *Red Deer Advocate* described the meal: cream of tomato soup, both roast turkey and roast pork with gravy, roast and creamed potatoes, Brussels sprouts and green peas, sage and onion stuffing, cranberry sauce, apple sauce, and rolls. For dessert, Christmas pudding was served with either a custard sauce or a brandy sauce, along with trifle and mince pies accompanied by fruit, cheese, and nuts.

was opened early in March 1943 at a ceremony bursting, according to the *Local Press*, with "exuberance of spirit and pride of accomplishment." The pride was not misplaced; they had raised over $6,000 to build and furnish the room, a bright and airy space where the men and women of the service could "pass the time on their hands amidst wholesome and congenial surroundings."

The city of Red Deer purchased an empty building and turned it into a recreation hut. The Knights of Columbus leased the building for $70 a month, and helped run it. The hut opened in December 1942, and was popular with service men and women from all three branches of the forces stationed near Red Deer. On average, 3,700 people made use of the facility each week, and on one Sunday in March the hut played host to 1,000 visitors. The weekly schedule of events included picture shows on Monday and Saturday nights, a dance on Thursday, bingo on Friday, and a concert featuring local talent on Sunday evening.

The men and women were welcomed into the community in other ways as well. Church groups, women's organizations, men's clubs, and others put on entertainments, held dances, organized dinners, and showed movies. Families took the airmen and airwomen into their homes for weekends, holiday dinners, and sometimes for short leaves. In Red Deer, servicemen got into the fair for 25¢–10¢ below the going rate. One airman recalled that, during his stay at the manning depot in Edmonton,

Staff and students from No 2 Wireless School take part in a parade through downtown Calgary. BCATP schools enthusiastically supported local activities. Courtesy Provincial Archives of Alberta: Harry Pollard Collection, P6494

it was never difficult to get to his home in a nearby small community; hitchhikers in uniform got rides immediately. The managing director of the High River station publicly thanked local residents for their support, cooperation, and friendship. The community, he said, had adopted each new class and helped to make the students' stay there happy ones.

British Commonwealth Air Training Plan stations contributed substantially to community life through the war years. The BCATP schools in Calgary regularly participated in the Stampede parade, often making humorous floats of painted aircraft. At every station, airmen and women threw themselves into victory loan drives, boosting community totals as well. In Red Deer in the spring of 1943, the Penhold station took part in a parade through town, the airmen marching along in uniform. In Lethbridge, pilots from the bombing and gunnery school "bombed" the city with leaflets to help start a war savings certificate and stamps drive. The stations held Christmas parties for local children, and helped communities raise money to buy items for the forces overseas.

Formation flying over the prairie. In many parts of Alberta, the drone of aircraft overhead was a constant reminder that a war was going on. Courtesy Glenbow Archives, Calgary, NA-4943-3

But the biggest contribution of the Air Training Plan to the communities it touched was economic. Survey crews used local facilities, and the subsequent construction boom saw local people hired and wages spent on food and housing and a hundred smaller things. As the station neared completion and staff began arriving, housing was in demand. Single men often lived on the base, but married men needed accommodation in the town. Local authorities helped by canvassing the population, trying to convince residents to rent rooms to the newcomers. Locals were also urged to undertake renovations that would make unoccupied real estate more appealing. Many families needed furnished accommodation, and that was hard to find. In October 1941, the *Lethbridge Herald* reported that, while many residents were taking in boarders and renters, the housing shortage in Vulcan remained acute. Some people were sleeping in garages, others were boarding with farmers, and an enterprising few were moving unoccupied houses from the surrounding district into Vulcan.

During construction of the Penhold station, the *Red Deer Advocate* reported that one man had turned his restaurant into a rooming house and gone to cook for the guards at the base. The Vulcan Board of Trade ran an "open letter to Vulcan householders" in the *Advocate*, explaining the housing demands soon to be made of the town. From Edmonton to Medicine Hat, home-owners with spare rooms could count on a regular extra income as a result of the Air Training Plan.

The stations brought a tremendous amount of money into nearby communities. In October 1941 the *Lethbridge Herald* carried a headline indicative of a school's impact: "Vulcan Booming as Air Station Brings Big Payroll." Vulcan was a "hustling centre," busier now even than it had been in the biggest years of the wheat boom. All businesses, from barbers to butchers, from grocers to garages to dry goods stores, were benefiting.

Many local people found employment at the stations as well, taking jobs that weren't restricted to military personnel. Men and women worked as mechanics, truck and tractor drivers, janitors, stenographers, and at a host of other occupations to keep the plan going. Their steady pay cheques helped keep the local economy buoyant.

Air stations and their personnel were welcomed in most communities, although there were some complaints of price gouging. One report in the *High River Times* noted that "in other towns things cost a lot more if a man is in uniform." People were urged to keep prices and rents fair. In some cases the Wartime Prices and Trade Board stepped in to make an adjustment, but overall there does not seem to have been any concerted effort to take advantage of service personnel.

While communities could easily get used to the extra money and the endless planes droning and zooming overhead, other aspects of having an air training station in your back yard were more difficult to accept. Crashes were vivid reminders that the world was at war. Some were caused by irresponsible flying, but most were the result of inexperienced pilots or unpredictable weather. When deaths occurred, the airmen were buried with full air force rites. After the death of a British pilot near Claresholm, the newspaper urged the public "to participate in this service as far as possible to pay tribute to a volunteer who died so far from home."

Sometimes the exuberance of young pilots took them stunting over the countryside or into a mock dogfight. Two brothers driving quietly along near High River were startled when a low-flying plane sliced the top off their car. They were cut by flying glass and had to walk several kilometres for medical assistance. The *Okotoks Review* carried an item about trainees from Calgary who were "performing stunts over Okotoks with a cheerful disregard for their own lives, those of others or the damage to their plane or other property that might result." According to the paper, the boys had looped the loop around the telephone wires and come alarmingly close to shaving the flag off the pole at the Hotel Willingdon. The mayor phoned the station requesting that these activities be put to a stop.

Eventually, all activities came to a stop. The acceleration and expansion of the plan in 1942 had led to a surplus of graduates, and by late 1943 there was a substantial reserve of trained men waiting for operational postings. In early 1944, schools began closing. Rumours of impending closures often circulated before any announcement was made, alarming town councils and boards of trade, who did not want to see an end to their economic boom. They saw the airfields as bridges to the modern world of transportation, and looked for assurances that they would remain operational in some capacity after the war. The High River Chamber of Commerce wrote the Minister of Defence for Air asking him to reconsider the decision to close the station there. They suggested that the school might offer refresher courses for air force personnel.

Despite the protests, the closures went ahead. On 14 September 1944 the softball team from the High River station was contending for the service league championship; on 16 November, the station had been closed. "It has all happened," the news-

WDs refuelling aircraft at No 2 Air Observer School in Edmonton. Ground crews made sure the planes were in top shape and ready to go. Courtesy City of Edmonton Archives, EA-10-3181.46.4

paper lamented, "with a suddenness that has left most people bewildered." Final wings parades were held across the province with the pomp and entertainments the public had come to expect. A notice in the *Vulcan Advocate* wished farewell and happy landings to the departing men and women. "Your conduct at all times in the town was exemplary," the notice read. "We enjoyed having you."

Sad farewells were said across Alberta, not just to the dollars that would be lost, but for friends and good times that might never be seen again. School officials, too, said their farewells. "It is difficult at times like these," the director of the High River station said prior to the school's closing, "to find words adequate with which to express our thanks and appreciation of the manner in which you have absorbed us into your district life and made us feel at home."

Some people stayed. Many a romance had taken flight in Alberta communities between school staff and trainees, and local residents. Some couples married and settled nearby. Others stayed for different reasons. High River instructor Jock Palmer, well known in Alberta aviation circles, stayed to set up an electrical repair shop in High River and hired a colleague from the EFTS.

Dismantling the plan was a mammoth task. Planes were salvaged, burned, or junked. Some were sold. The Crown Assets Allocation Committee was in charge of disposing of the buildings and supplies. An auction sale at High River lasted fourteen hours and attracted over 2,000 people. Everything from "a 30 foot [9.144-metre] hardwood veneer liquor bar with cut glass shelves" to tools and garage equipment, theatrical spot lights, meat-cutting blocks, and public address systems was sold. Many goods collected by the communities for the stations were also sold at auction. At Vulcan, for instance, a piano, electric clock, Victrola, rugs, irons, thirty-five pairs of curtains, two leather-covered hassocks, gas lamps, and "other articles too numerous to mention," as the advertisement said, went to the highest bidders at a giant auction on 9 June 1945. The proceeds went to the Red Cross.

"Yanks Come thro' By Air and Land"

The constant drone of BCATP planes overhead was not the only reminder that the world was at war. North American defence plans had always operated under the assumption that any aerial attack would come from the Atlantic. Following the Japanese attack on Pearl Harbour in December 1941, Allied governments had to acknowledge the possibility that an attack might originate in the east. Alaska seemed particularly vulnerable, its long arm extending across the Bering Sea, with little muscle behind it to repel an attack. One way to get troops, supplies, and planes to Alaska's defence was through northwestern Canada.

Rudimentary facilities already existed in northern Alberta, thanks to mineral exploration and development. The Canadian government had also surveyed an aerial route from Alberta through British Columbia, the Yukon, and the Northwest Territories in 1935, hoping it might be part of a great circle route to the Orient. Yukon Southern proved the viability of the route when it established a mail run from Edmonton through Fort St John, Fort Nelson, and Watson Lake to Whitehorse in 1937.

In 1939, the Department of National Defence decided to continue developing this route, and began to upgrade facilities at Grande Prairie as well as the stops along Yukon Southern's route. Before hostilities changed the face of this aerial highway, the improved route had been intended to speed exploration and development, and expand northern mail service.

Construction continued slowly after the outbreak of war, as the situation in Europe demanded military resources be focused there. Encouraged by the Canada–United States Permanent Joint Board on Defence, and a gradually increasing unease over rising tensions in the Far East, the building did continue. By September 1941, most of the fields, including the one at Grande Prairie, were ready for daytime use. By December, with radio range equipment installed, night flying could begin. One man recalls flying the route with Yukon Southern to take up his position as head of the radio range station in Whitehorse in 1941: the flight from Edmonton to Grande Prairie was completed in a Barkley-Grow with twelve passengers in seats lining the fuselage. The baggage and freight were piled in the aisle running down the middle.

The United States had begun improving and expanding airfields in Alaska, completing the line of landing fields that now ran from Edmonton to Fairbanks. Called the Northwest Staging Route, it permitted aircraft to get to Alaska fairly quickly. More important—for the Japanese threat to Alaska soon passed—it was the route taken by planes destined for Russia under the lend-lease program. American bombers, transport planes, and fighters were flown to Alaska by American pilots. Russian pilots then picked them up and flew them the rest of the way. For Grande Prairie, it

Yukon Southern's Barkley-Grow, the Yukon Queen, at Grande Prairie's airport in 1941. Courtesy Provincial Archives of Alberta: Dan Campbell Collection, C162

meant a tremendous change to the airfield. A 900-metre runway was added, and American military planes roared constantly overhead. Hundreds of military personnel moved in.

The United States began sending planes along the route in early 1942. Accidents and crashes among the early flights pointed up the need for more emergency landing fields and better communications systems. As well, the airfields had to be continually upgraded to land the heavier planes. Japanese activity in the Aleutians in June 1942 spurred the Americans to rush the completion of the airfields, sending every available plane along the route delivering supplies. Lethbridge and Calgary both saw increased traffic because of the Northwest Staging Route. But it was Edmonton, a major landing point on the route, that saw the greatest increase.

The number of planes travelling the route would grow to staggering numbers, and Edmonton was once again called upon to be a gateway to the north. Edmonton had offered its airport to the federal government immediately when war broke out. The offer was accepted, as it was at civilian airports across the country. Immediate improvements were made to the runways and taxi strips. Already busy with commitments to the British Commonwealth Air Training Plan, Edmonton grew far busier when the Northwest Staging Route swung into action and other defence projects got underway. American personnel moved into the city, and the planes kept coming.

Congestion at Blatchford Field, June 1942. During the war Edmonton's airport became an important stop on the American ferry command route. Planes being flown to Alaska for delivery to the Russian air force under the lend-lease program all stopped in Edmonton. As well as BCATP traffic, and flights connected with projects such as the Alaska Highway, the addition of lend-lease planes made Edmonton's airport one of the busiest on the continent. Courtesy Provincial Archives of Alberta, A 5300

Hangars and other buildings went up. Runways were lengthened and improved. A new administration building opened in 1942 to handle the increased flow of traffic and to provide services to air travellers. Commercial airlines were also taking part in the effort in the skies, putting another strain on airport resources.

The skies over Edmonton were a blur of planes heading for Russia. Planes were squeezed into every available corner of the airfield. If weather caused a backup in delivery, the congestion grew worse. In one twenty-four-hour period in June 1942, 500 planes touched down in Edmonton. On 29 September 1943, over 850 arrivals and departures were noted in the airport log.

There were other planes in Edmonton's skies. Leigh Brintnell had formed an aircraft maintenance organization with two other airlines and started repairing planes in 1936. When World War II erupted, Aircraft Repair Limited was ready. Its first job was repairing and reconditioning planes that had taken part in aerial battles in Europe. But soon damaged planes from the lend-lease program and the BCATP were limping into the repair shop. Keeping all those planes in the air was a colossal undertaking. In April 1941, the first twenty-five employees of Aircraft Repair Limited moved from No 2 hangar at the airport into their own plant and began work. By March 1943, with a staff of more than 1,750, Aircraft Repair claimed to be the largest overhaul plant in Canada, handling everything from entire airplanes to engines, radiators, and instruments. In October of that year, the men and women of Aircraft Repair celebrated the overhauling of their 1,000th plane and 1,500th engine. The company also had a field crew who travelled to BCATP stations in Alberta and Saskatchewan repairing aircraft that had sustained minor damage in crashes or mishaps.

Brintnell took the responsibility of aircraft repair very seriously, realizing that his operation was playing a critical role in the war effort. An article in the *Edmonton Journal* described a large, efficient operation that involved two main assembly depots, smaller warehouse buildings, a hangar, two engine-testing stands, and a cafeteria. Each worker had a personal set of tools, while precision tools that were used only occasionally were kept in a central depot and signed out when needed. Prizes were regularly given to employees who came up with time-saving ideas, and a full-time coordinator was assigned to organize morale-boosting activities aimed at keeping the work force happy and productive.

While most of its business was hauled in, a lot of it flew out. Archie McMullen was a test pilot for Aircraft Repair Limited, and his log books show hundreds of flights on a wide variety of planes, including Fairey Battles, Ansons, Harvards, Norsemans, Oxfords, Cessnas, Lodestars, Bolingbrokes, and a few Yales.

Blatchford Field also handled air traffic that had been generated by other aspects of the war effort. The Alaska Highway, the land route to Alaska, was pushed forward with dizzying speed by American engineers and troops. The route ran from Dawson Creek, British Columbia, through Whitehorse and on to Alaska. Supplies and personnel began arriving in Edmonton in March and April 1942 before being moved to their postings. "Yanks Come thro' By Air and Land" the *High River Times* reported in June, mentioning in particular the roar of plane engines heard throughout the night. Construction began simultaneously at several points. Airports and landing fields, including those at Edmonton, Fort McMurray, and Grande Prairie, all experienced increased activity as planes ferried people and supplies northward. Although the highway essentially followed the Northwest Staging Route, crude landing fields were slashed out of the bush when and where they were needed. Pilots flying the Northwest Staging Route commented on how helpful the Alaska Highway was in keeping them on course: if the weather cooperated, they could simply follow the gravel scar all the way to Alaska.

Additional air traffic was created by the Canol Pipeline, another joint project in support of the war effort. Designed to increase the supply of gasoline and other petroleum products to defence undertakings in the northwest, the plan involved increasing production at Norman Wells, building a pipeline to carry the crude oil to a new refinery to be constructed in Whitehorse, and building additional pipelines to distribute the refined products. The landing strip at Fort McMurray was enlarged, while Peace River, Embarras, Grande Prairie, and Calgary all saw their air traffic increase as a result of defence construction in the northwest.

So great was the pressure on Edmonton's airport, and so large was the American presence, that another airfield was necessary. Construction of Namao airfield was begun in the summer of 1943. Planes began using the field in 1944, taking off from the two massive concrete runways. Namao operated as an American base until the end of the war.

One of the largest demands put on municipalities was for power. In Calgary, street lighting had to be improved because of the BCATP activities in the city, including the

The cramped facilities at Aircraft Repair Ltd. Industries involved with aviation surrounded the Edmonton airport. Courtesy City of Edmonton Archives, EA–10–3181.47.6

lighting on Edmonton Trail leading to the municipal airfield. Power lines had to be extended to the radio transmitters, too, and other work was undertaken for the Department of Transport. The responsibility for such things often fell on city shoulders, although the Department of Transport was billed for some of the work. The annual report for Calgary's Municipal Airport for the year ending 31 December 1941 noted that, along with improved gas, water, and sewage connections and a paved parking lot, the garden plots looked "exceptionally nice." "Very little private flying is done at the Airport now," the report concluded. "Military training predominates."

The report covering 1942 had an entirely different tone: there was no mention of gardens. The Department of Transport now had control of the airport, and staffed it with its own personnel and equipment. The runways had been hard surfaced, taxi strips had been added, and lighting and drainage had been improved. The city's only interest remained in the hangar, and that was leased to Northwest Ferry Command. Additional improvements and expansion continued throughout the war.

Cities participated in the aerial war in other ways. Calgary and Edmonton each "adopted" operational squadrons. Calgarians took on the Wolf squadron; the city received adoption papers, and each plane in the unit carried the name "City of Calgary." Citizens undertook to raise money to send cigarettes, games, and other comforts to the ground and flying crews of the squadron. Chocolates, tins of coffee, and hard-to-come-by necessities such as razor blades were greatly appreciated by military personnel overseas. The men in turn sent letters of thanks, and sometimes pictures of contented pilots and navigators puffing on Brier pipe tobacco or MacDonald's Menthols. Cities and towns also took part in campaigns to purchase aircraft. Edmontonians threw themselves into the Spitfire campaign in late 1940. A diamond ring was donated for a raffle, women sold jellies and jams, boy scouts raised money, and bridge and whist parties all combined to bring in more money than was needed.

The housing shortage that plagued smaller centres near BCATP installations was felt in the larger cities, too. Both Edmonton and Calgary pleaded with their citizens to open their homes to the newcomers. Many west-end Edmonton residents rented rooms to military personnel for $25 a month. When No 4 Training Command of the RCAF moved from Regina to Calgary, accommodation had to be found for over 500 people. As personnel connected with developments in the northwest arrived, or commercial companies assisting in the war effort expanded their staffs, new pressures were placed on the housing market. Aircraft Repair Limited also called for something to be done: its work would suffer if adequate housing could not be found for its workers.

As in the smaller centres, the economies of the larger cities grew as the military presence increased. Supplies and provisions bought in Edmonton were shipped north, while souvenirs and other goods purchased by American soldiers were shipped south. Dance halls, bars, and stores of all kinds saw their cash registers swell with wartime pay. From taxis to storage space, everything was hard to come by.

Some of the crush was caused by the operations of commercial companies. In Lethbridge, the energetic Aviation Committee of the Board of Trade was successful in adding that city to Western Air Express's Los Angeles–Great Falls route. Trans-

Canada Air Lines continued operating out of airports in Alberta, and showed an increase in mail and passengers carried, much of it owing to the military situation. After 1942, though, TCA had a competitor.

By 1940, many of the small companies operating in various parts of the country could not make ends meet. Most of them lacked capital, were flying inefficient planes, and were trying to elbow their way along crowded routes. Factories that once produced spare parts for their planes now made only wartime commodities. Even Yukon Southern, an airline that had several planes and scheduled routes between Edmonton, Vancouver, and the Yukon, was in financial trouble.

The Canadian Pacific Railway had been eying these small regional operators, and in late 1940 began buying them up. By the end of 1941, the CPR had purchased ten companies whose individual operations, when added together, almost spanned the country. In July 1942, Canadian Pacific Air Lines Limited began to fly. Among the transportation giant's purchases had been Yukon Southern Air Transport, Mackenzie Air Service, and Canadian Airways. These companies had carried out a vital support role during the Alaska Highway and Canol projects, transporting aerial survey crews and load after load of supplies. This role continued after the formation of Canadian Pacific. Grant McConachie stayed on with the new company, but continued in his old ways, using the same persuasive techniques that had beguiled banks and oil companies and kept Yukon Southern in the air, to snatch brand new Lockheed Lodestars right from under the nose of the war effort. Canadian Pacific Air Lines

Planes at Fort McMurray, 1942. Courtesy Provincial Archives of Alberta, A7652

estimated that 90 percent of its traffic was war-related in 1943, but civilian flights did continue along scheduled routes. One or two new routes were even added.

By 1944, military activities were beginning to wind down. BCATP facilities were closing, American personnel were pulling out, and the flow of supplies and planes through Alberta to the northwest was slowing down. Aircraft Repair Limited began letting staff go as business dwindled. As early as 1944, city officials had begun to ask what would happen to their greatly expanded airports when the war ended. In Calgary, people wondered whether the federal government would continue to run the airport, or whether control would revert to the city.

Then the war was over. First in Europe, then in the Pacific, the carnage stopped, and joyful celebrations took over. But though the streets were full of celebrants, Alberta's skies were quiet.

Inside the control tower at the Edmonton airport, April 1944. Courtesy Provincial Archives of Alberta: Alfred Blyth Collection, Bl. 714/2

One Day Wide

Trans-Canada Air Lines had declared Canada to be only one day wide in early 1943. TCA offered daily transcontinental service and international flights, as well as inter-city service. It was preparing for peace and reminding the population that flight would be an important part of postwar life. Canadian Pacific was not standing quietly by. It was reminding Canadians of its important role in opening the north, and how that role would expand after the war.

In September 1944, a group of people involved with aviation from private companies, government, and communities in southern Alberta and the United States gathered at the Palliser Hotel in Calgary for the Southern Alberta Aviation Conference. The delegates were treated to a largely optimistic view of postwar aviation that included fantastic increases in air traffic linking their communities with the rest of the world. An increase in the number of private planes was also predicted. There was a good deal of discussion of airport facilities and the requirements of postwar aircraft.

The sentiments expressed in Calgary were echoed across the country. There was no doubt that World War II had pushed Canada into the modern aviation age. Plane travel had become fast and frequent, and trans-oceanic travel had been seen to be possible. Civil aviation was poised to take off on a grand scale around the world. Canada responded by amending the *Aeronautics Act* in 1944 to permit the creation of a three-member Air Transport Board to advise the Minister of Transport on civil aviation issues. Determining routes, licensing airlines, and discussing regulations were all part of its mandate. Canada also played an important role in determining a course for international civil aviation in a conference in Chicago in 1944.

Civil aviation in Canada had emerged from the war dominated by national carriers, as opposed to the largely regional system it had been in 1939. The two national airlines provided the bulk of the service, with a few small regional outfits emerging to service local needs. Alberta cities, encouraged by the modern aviation infrastructure the war had established across the province, jockeyed for position in the new national framework. From Peace River through Grande Prairie to Fort McMurray,

Edmonton, Calgary, and Lethbridge, and from Medicine Hat east and west to both provincial borders, the war had left a legacy of runways, hangars, radio beacons, and weather forecasting facilities. Flight had come to represent modernity, and with the winding down of the war effort, many communities had launched campaigns to maintain their facilities and convert them to peacetime uses.

Both Calgary and Medicine Hat set up aviation committees to ease the transition. In Calgary the Special Aviation Commission advised municipal authorities on the development, operation, and promotion of the city airport. The Commission sought answers from the Department of Transport concerning the postwar use of air facilities. The runways in Calgary were not heavy enough to support large planes. The federal meteorological service had been taking readings in the area for a couple of years, lending some support to the theory that a new site was being selected. Given that, Calgary was reluctant to put money into improvements on the current site. Flights using heavy aircraft were moved to Lethbridge pending runway improvements, and this did not please Calgary's civic authorities.

Calgary was also becoming aware of how expensive running an airport could be. Although the city had bought the land, put up a hangar, and built some roads in 1929, a good deal of the subsequent development had been at the expense of the federal government. The city was not only looking at upgrading the facilities,

A thoroughly modern Trans-Canada Air Lines plane being refuelled in Edmonton.
Courtesy City of Edmonton Archives, EA-160-1374

but also at the substantial upkeep that lighting, improved runways, taxi strips, and drainage would require.

The commission's report for 1947, however, showed a turn-around, with optimism replacing the initial confusion and unease. The commissioners declared the airfield to be one of the leading light-plane fields in Canada. The Department of Transport had extended the north-south runway, making it big enough for the largest planes operating from Calgary. TCA had put Calgary on its transcontinental route, and had improved and enlarged its office and facilities as well. Lighting at the airport was being improved, and what the report described as the "latest Instrument Landing System" was being installed. The airport had recorded 52,630 landings and takeoffs in 1947. On 1 July 1949, the city assumed control of the airport under an agreement with the Department of Transport. The department would still be responsible for building and repairing taxi strips, sealing the runways, and improving the lighting.

Lethbridge also enjoyed a favoured position in immediate postwar aviation development. For a few years after the war, eighteen TCA flights and two from an American airline landed every day.

The victorious Edmonton Mercurys hockey team back in Canada after winning the 1950 world hockey championships in London, England. The team travelled across Canada by plane, but crossed the Atlantic on an ocean liner. Courtesy Jack Manson

Cities and towns in Alberta soon discovered, however, that geography and technology played as large a role in postwar aviation development as they had in earlier years. All the civic boosting possible could not keep Lethbridge a major east-west landing location when the Rockies were no longer a barrier, and planes did not have to depend on the Crowsnest Pass as a thoroughfare. Medicine Hat, also lobbying to become a destination on national flights, canvassed local businesses to learn how often they would use such a service if TCA landed in the city. Medicine Hat did get on TCA's schedule for a while, a fact authorities attributed in part to their survey results. Red Deer, also hoping to become a TCA destination, was considering a survey to bolster its argument. But as planes got bigger and faster, and the physical environment became less of an obstacle, smaller centres lost their privileged aviation status.

Regional air operations found it difficult to be profitable right after the war. Although a few companies tried to get going, by 1950 there were only a couple left. In northern Alberta, Grande Prairie and Peace River rekindled their hopes that they might be able to siphon some of the northern traffic away from Edmonton. Small companies such as Yellowknife Airways and Northern Flights did serve parts of northern Alberta for a few years, but the Peace River–Grande Prairie area could not

The Flying Farmers, an organization begun in the United States toward the end of World War II, quickly spread to Canada. In this photo, a group has just landed in Edmonton. Courtesy Provincial Archives of Alberta: Dan Campbell Collection, C364

sustain the aerial importance it had enjoyed during the war. Things were not much better in the south. In 1950, Medicine Hat lost the charter company that had been operating in the area. While the aviation committee lamented the general loss of flights, it was particularly concerned that no air transport would be available should a medical emergency requiring quick transportation arise.

The increase in the number of planes in private and business hands may have been a factor in the demise of local charter companies. When Medicine Hat authorities contacted the provincial government about the possibility of a provincial air ambulance service, the reply was in the negative. Private planes were always made available in emergencies, the Minister of Industries and Labour maintained, making a provincial service unnecessary.

Some regional carriers did make quite a success of it. Thomas Payne "Tommy" Fox, an air force veteran, started Associated Airways in Edmonton right after the war with two war-surplus aircraft. Associated initially concentrated on passenger carrying and pilot training. By 1948, the company was operating across the north, soon becoming an important cargo hauler.

Recreational and business flying were gaining in popularity all over the prov-

The airplane fitted right into the post-war oil boom, and growth in the oil industry spurred an increase in flying. Courtesy Provincial Archives of Alberta: Harry Pollard Collection, P3237

ince. More and more enthusiasts were purchasing their own small planes and getting their private pilot's licenses. Flying clubs regrouped after the war, and were soon busy again, flying and teaching. Many used aircraft such as the Tiger Moths that had been declared surplus at the end of the war. The federal government and the RCAF once again supported flying club activities as a way of ensuring that Canada had a reserve of trained aviators.

In business circles, companies were recognizing the advantages of air travel, a point the airlines stressed. "Business trips by TCA are economical," an advertisement in the December 1947 *Lethbridge Herald* declared, adding, "Especially to the business executive, time is money." The oil boom precipitated by the Leduc discovery in 1947 resulted in greatly increased air traffic as executives hopped from head office to their wells by air. Not only were businesses discovering the ease of airline travel, many were also enjoying the convenience of a company plane. From the Hudson's Bay Company to oil, gas, and pipeline firms, company planes were becoming common. The use of planes in agriculture was growing, too. Some ranchers found they could keep an eye on their herds from the sky, and a couple of companies, including Westland Spraying Service Limited in Airdrie, tried to make a go of it with crop dusting.

Although airline companies had taken on an almost military role during the war, aviation came out of the war with civil and military responsibilities clearly delineated. The role of the air force in national and international security was now entrenched in national policy. Airline companies carried people and parcels, while the RCAF looked after military matters, and search and rescue, which had become part of the military's responsibilities during the war. Wop May had set up a search-and-rescue unit when he was running No 2 Air Observer School, believing that better search and rescue capabilities would be needed on the Northwest Staging Route. May had become an aviation pioneer once again when he established an RCAF parachute school in Edmonton in 1944 as part of the Northwest Air Command headquartered in Edmonton. Edmonton, maintaining its role as gateway to the north, became western headquarters for the RCAF's postwar search-and-rescue operations.

Edmonton remained a busy air harbour as traffic northward and on the transcontinental system assumed postwar patterns. Municipal authorities resumed control of the airport on 1 November 1946, although the details of the agreement with the Department of Transport were not worked out until some time later. Edmonton continued to be a national leader in air traffic volume. Steady growth in business and recreational flying after the war, as well as the continuing evolution of the national carriers, saw activity at the airport increase yearly. Over 81,000 landings and takeoffs were logged in 1948. When the Department of Transport established regional offices in 1948, one was set up in Edmonton. When the RCMP re-established its Air Service in 1946, a plane was based in Edmonton. The 418 squadron made its home in Edmonton when it returned from Europe in 1946, staying until it moved to Namao in the 1950s.

The ever-resourceful Leigh Brintnell continued to put planes into the skies around Edmonton for a few years after the war. Now president of Northwest Industries, his

Leigh Brintnell's manufacturing and repair company underwent another transformation with the end of the war. Emerging as Northwest Industries, the company turned out Bellanca 31-55 Senior Skyrockets. The photograph above shows part of the assembly progression, with the wings, fuselage, and motor being added to the frame. The photo below shows an almost completed plane nearing the end of assembly. The plane in the left foreground, CF-DCH, was the first to come off the assembly line. Courtesy City of Edmonton Archives, EA-340-120, and Provincial Archives of Alberta, A5884

new company manufactured Bellanca 31-55 Senior Skyrockets under license from the Bellanca Aircraft Corporation of Newcastle, Delaware. The first Skyrocket to emerge from the plant, CF-DCH, was taken on its first flight on 28 February 1946. The plant shut down so all the employees could watch the test flight. For half an hour the green and yellow plane circled the Edmonton airport. Only thirteen Skyrockets were built at Northwest Industries, however. Production stopped in 1949 as the Skyrocket became a victim of its disadvantages—its wood and fabric construction—and the appearance of de Havilland's popular all-metal Beaver.

With increased international traffic, and bigger planes on all routes, municipal airports that had once been located in empty fields but were now engulfed by urban development were coming under pressure. Jimmy Bell, still the manager of Edmonton's municipal airport in 1949, confessed to *Maclean's* magazine that he often wondered how he could maintain an international airport five minutes from the "castle-like Macdonald Hotel." Early in the 1950s, Edmonton broke ground for a new international airport outside the city.

By 1950, Alberta was well integrated into national and international air patterns. In the spring of 1950, crowds of curious people had gathered at the airports in Lethbridge, Calgary, and Edmonton to welcome the inaugural flights of Northwest Orient Airlines and Western Airlines. Now, Alberta would be regularly linked by air to cities all across the world. Northwest's plane was christened the "Province of Alberta" in front of a cheering crowd in Edmonton.

So many other flyers, from Eugene Ely to Katherine Stinson, from Fred McCall to Wiley Post, and from Ernest Hoy to all the fresh-faced British Commonwealth Air Training Plan graduates, had been cheered by similar crowds across the province. Those crowds had marvelled at their achievements, and congratulated the participants. They had watched as fragile open-cockpit, linen-winged airplanes were gradually replaced with sturdy, enclosed, all-metal aircraft. They had witnessed the world open up as their own province shrank, and seen the distance you could travel in a day come to be measured in countries rather than kilometres. When the bottle that had been filled with water from every stop on Northwest's flight was broken over the plane's nose at the Edmonton airport, the spectators knew that now, for them, much of the world was one day wide.

Sky riders had captured Alberta's heart once again.

References

There is an abundance of material available on the history of aviation in Canada, and Alberta; I will not list everything here. What I will try to do is list books, articles, journals, and manuscript collections that have been helpful in the preparation of this narrative. From this compilation, both the general reader and the specialist may find useful material and guideposts to the literature in general. A more detailed chapter-by-chapter discussion follows the general discussion. Items mentioned more than once will appear in abbreviated form in subsequent references.

There are many works that deal with the early history of aviation throughout the world. The many books of C. H. Gibbs-Smith are particularly useful. See his *Flight Through the Ages* (London, 1974) and *Aviation: An Historical Survey from its Origins to the End of World War II* (London, 1970). See also Walter J. Boyne, *The Smithsonian Book of Flight* (Washington and New York, 1987) and *The Lore of Flight* (New York, 1974); J. L. Naylor and E. Ower, *Aviation: Its Technical Development* (London, 1965); Alvin M. Josephy, Jr, ed, *The American Heritage History of Flight* (1962); Frank Howard and Bill Gunston, *The Conquest of the Air* (London, 1972); and Clive Hart, *The Dream of Flight: Aeronautics from Classical Times to the Renaissance* (London, 1972).

Canada has several aviation histories. Frank H. Ellis's contributions are very useful. See his *Canada's Flying Heritage* (Toronto, 1954) and *In Canadian Skies* (Toronto, 1959). Ellis's *Canadian Civil Aircraft Register,* published by the Canadian Aviation Historical Society, is an exhaustive listing of aircraft registered in Canada to 1945. J. R. K. Main's *Voyageurs of the Air: A History of Civil Aviation in Canada, 1858–1967* (Ottawa, 1967) and Larry Milberry's *Aviation in Canada* (Toronto, 1979) are good overviews of the history of flight in Canada. G. A. Fuller's *125 Years of Canadian Aeronautics: A Chronology 1840–1965* (Willowdale, 1983) is a useful place to start, although it is less comprehensive for Alberta than for other parts of the country. David MacKenzie's *Canada and International Civil Aviation, 1932–1948* (Toronto, 1898) is an excellent discussion of Canada's role in aviation on the international stage. Margaret Mattson's "The Growth and Protection of Canadian Civil Commercial Aviation, 1918–1930" (PhD thesis, University of Western Ontario, 1978), and William McAndrew, "The Evolution of Canadian Aviation Policy following the First World War" in *The Journal of Canadian Studies* (Vol 16, nos 3 & 4), 86–99, should also be consulted. General works on aviation personalities include Shirley Render, *No Place for a Lady: The Story of Canadian Women Pilots, 1928–1992* (Winnipeg, 1992), and Alice Gibson Sutherland, *Canada's Aviation Pioneers: 50 Years of McKee Trophy Winners* (Toronto, 1978). *Canada's Aviation Hall of Fame* (1990) gives brief biographies of each inductee to 1990. Another useful general overview is Lorne Manchester's *Canada's Aviation Industry* (Toronto, 1968). See also K. M. Molson, *Canada's National Aviation Museum: Its History and Collections* (Ottawa, 1988);

and K. M. Molson and H. Taylor, *Canadian Aircraft Since 1909* (Stittsville, ON, 1982). J. A. Foster's *For Love and Glory: A Pictorial History of Canada's Air Forces* (Toronto, 1989) is useful for a general outline.

For Alberta, readers should consult Stanley Gordon, *The History of Aviation in Alberta to 1955* (Background Paper 25, Reynolds–Alberta Museum, 1985). For Edmonton, Eugenie Louise Myles's *Airborne From Edmonton* (Toronto, 1959) contains many useful details. Bruce Gowans has made a tremendous contribution to the history of flight in Alberta with his painstakingly researched *Wings Over Lethbridge* (Occasional Paper No 13, Whoop-Up Country Chapter, Historical Society of Alberta, Lethbridge, 1986) and *Wings Over Calgary* (Chinook County Chapter, Historical Society of Alberta, 1989). Henderson's *Gazetteer* and *Directories*, and Wrigley's *Directory* are useful to trace the activities of businesses and people.

Newspapers were extremely valuable sources for this study. Aviation events and developments in Alberta and around the world were regularly reported to a public hungry for news of this new technology. Newspapers researched for this study were, in alphabetical order, *Banff Crag & Canyon, Calgary Herald, Claresholm Local Press, Coleman Journal, Edmonton Bulletin, Edmonton Capital, Edmonton Journal, Financial Post, Grande Prairie Herald, High River Times, Lethbridge Herald, Medicine Hat News, Northern Gazette* (Peace River), *Northern Record* (Fairview), *Okotoks Review, Peace River Record, Red Deer Advocate, Vulcan Advocate,* and *Wetaskiwin Times.*

Several collections held by archives in Alberta were consulted, and they are listed below. Most of these archives also have oral history collections that contain material on aviation.

Provincial Archives of Alberta, Edmonton (PAA)

Aeronautics–Miscellaneous Clippings Acc.70.38
Scrapbook Acc.71.102
Photographs Acc.73.549
Manuscript Acc.75.507
A. M. Berry Papers
Robert Torrance Papers
Leigh Brintnell Papers
Archie McMullen Papers
Wop May Papers
Gladys Walker Papers
James A. Bell Papers

Glenbow Archives, Calgary

William Donald Albright Papers M8
Calgary Exhibition and Stampede Papers M2160
Canadian Pacific Railway Records M2269
Major Edwin R. Carr Papers M3602

Warren Clarke Papers M3725
Katherine and David Coutts Papers M278
G. A. Fuller Papers M3726
Elmer Fullerton Papers M7106
John Gladstone Papers M3727
Alfred Goodwin Papers M441
Stanley N. Green Papers M454
Dr Harold A. Hamman Papers M2278
J. D. Higinbotham Papers M517
Wilfrid Reid "Wop" May Papers M829
Fred Robert Gordon McCall Papers M716
W. Roland Murray Papers M4113
John E. Palmer Papers M931
Joseph Patton Papers M946
Elizabeth Bailey Price Papers M1000
Ross Family Papers M6176
William E. Rowbotham Papers M1076
George William Scott Papers M1112
Transportation Oversize File M2624
Reverend Stephen M. Wik Papers M4712

Glenbow Library, Calgary

Edmonton Air Show Official Directory, 17 Sep 1930
Mackenzie Air Service Schedules Tariffs, 1 Dec 1939
Souvenir Programme No 5 EFTS Third Anniversary, 4 Sep 1944

Canada's Aviation Hall of Fame, Wetaskiwin

Inductees' Files

City of Calgary Archives, Calgary

Annual Reports
City Clerk's Papers
Special Aviation Commission Papers, 1946–1951

City of Edmonton Archives, Edmonton

Papers of the City Clerk R.G.8
Papers of the City Commissioners R.G. 11
Edmonton Exhibition Association Ltd Collection MS 322
Bush Pilots Clipping Files
Edmonton Municipal Airport Clipping File
EA-10-3181 Golden Anniversary of Powered Flight Exhibit Panels

City of Lethbridge Archives and Records Management, Lethbridge
Air Regulations, Air Board, Canada, 1931 P19851047035
Aviation History Reference Notes, Papers P19911057126, P19911057127, P19911057128
Chamber of Commerce Papers P19911057123
Alex Johnston Papers P19770239002, P19891016053, P19911016054
Lethbridge Aircraft Papers P19891068006

Medicine Hat Museum and Art Gallery Archives, Medicine Hat
Chamber of Commerce Collection M86.9
No 34 Service Flying Training School Assorted Papers M83.4.2, M83.4.4, M84.59.6
Clippings Files
    Military—World War II and Service Flying Training School
M78.31.4 file 4
"No 34 SFTS" M85.2.1
Air Mail M86.13.1

City of Wetaskiwin Archives, Wetaskiwin
Photograph Collections
Newspaper Clippings Files-93.8YE, Airport, Clubs and Societies
Collections/Depositors-Don MacEachern; Siding 16; Stan Reynolds; Peter Langelle; Biddy Odell

University of Alberta Archives, Edmonton
William Rowan Papers

Red Deer and District Archives, Red Deer
Clippings Files-Red Deer Airport, Air Travel, Penhold Air Base (Courses and Training, Special Events, Buildings and Facilities, and History and Anniversaries)
George Kerr Donation
The Penhold Log, Dec 1941 and Aug 1943
Photograph Collections
    E. Wells File
    E. Timble Album and Scrapbook

Fort McMurray Historical Society, Fort McMurray
Photograph Collection

Miscellaneous Collections
Stanley G. Reynolds Personal Collection
Stanley Gordon Research Files
Keith R. Spencer, Edmonton

Gas Bags and Bird Men

The chronology of these early events in Alberta, usually accompanied by fascinating observation and commentary, can be followed in the appropriate newspapers.

For expanded discussions of the idea of flight leading to the triumph of the Wright brothers, see the general works mentioned previously. Their experiments can be followed in Marvin W. McFarland, ed, *The Papers of Wilbur and Orville Wright*, Vol One, 1899–1905 (New York, 1953).

On balloons and airships, see Ellis, *Heritage*, chapter 7; and David C. Jones, *Midways, Judges and Smooth-tongued Fakirs* (Saskatoon, 1953), 1–4. The account of the appearance of the airship in Calgary in 1908 in the Calgary Exhibition and Stampede Papers at the Glenbow Archives captures the excitement the event created.

On early experimenters in Alberta, see Gowans, *Calgary*; Ellis, *Heritage*, chapters 2, 3, and 5; Ellis, *Skies*, chapter 2; PAA, Acc 75.507, "Canada's First Flying Machine"; and Hugh Dempsey, "Early Flying Machines," *Glenbow Magazine* (May/June 1986), 12–14. Reginald Hunt has been the source of some controversy: not all aviation historians are convinced he made his 1909 flight. As well as the contemporary newspaper, see Myles, *Airborne*, 12–13, and the *Edmonton Journal*, 2 Aug 1979, "Pioneer aviator was a forgotten hero." See also Hugh Dempsey, "The Ups and Downs of Alberta's First Aviators," in *The Boom and the Bust*, Alberta History, Vol III, Ted Byfield, ed (Edmonton, 1994), 53–4.

On the Aerial Experiment Association and its members, see Milberry, *Aviation*, 12–14; J. A. Foster, *For Love and Glory: A Pictorial History of Canada's Air Forces* (Toronto, 1989), 9–10; "John A. D. McCurdy," in Sutherland, *Pioneers*, 253–59; J. H. Parkin, *Bell and Baldwin* (Toronto, 1964), especially pages 40–166. See, as well, H. Gordon Green, *The Silver Dart* (Fredericton, NB, 1959).

Copies of patent records from the United States Patent Office for Newbury, Kelsey, and Hendrickson can be found at the Glenbow Archives, M2305.

The newspapers are again the best source for exhibition flying. See, as well, Gowans, *Calgary* and *Lethbridge*; and Myles, *Airborne*. An original poster advertising Eugene Ely's flight in Lethbridge is held by the Glenbow Archives, M2624. For Katherine Stinson, see also Myles, *Airborne*, 19–26; and Claudia Oakes, *United States Women in Aviation through World War I*, Smithsonian Studies in Air and Space, Number 2 (Washington, 1978), 33–36.

"Speed . . . and a Clear Highway in the Air"

For the debate on Bill 80, see the official *Report* of the Debates of the House of Commons of the Dominion of Canada, 1 April–5 May 1919, Vol 2, 1864–65, 2050–61. For the Air Board Act, see *Acts* for the Parliament of the Dominion of Canada, 20 Feb to 7 July 1919, Vol 1, Public General Acts, 31–34. For the 1920 Air Regulations, see *Supplement* to *The Canada Gazette*, Saturday, Jan 17, 1920, Orders in Council, #2596, 1–21. See *The Canada Gazette*, 24 Jan 1920, 2324–26 for position descriptions as advertised for Air Board employees.

For early post-World War I companies, see Gowans, *Calgary* and *Lethbridge*; Myles, *Airborne*; and the newspapers. See also Harry Fitzsimmons, "Barnstorming

Days," *Alberta Historical Review* (Spring 1970), 18–31. On Captain Fred McCall, see, as well, the Fred Robert Gordon McCall Papers, M716, at the Glenbow Archives; and William R. Murray, "The Fifth Ace: The Biography of Capt. Fred R. McCall" (unpublished manuscript, 1969), also at the Glenbow.

On the Red Deer fairs, see Westerner Exhibition Association, *100 Years of Progress: A History of Red Deer Fairs and Exhibitions* (1991). The Wop May Papers at the PAA and the Wilfrid Reid "Wop" May Papers (M829) at the Glenbow contain some details about May's early companies. See, for example, the statement of assets and profit and loss accounts for the immediate postwar years in M829. May–Gorman Airplanes tried to get one of the war surplus hangars but were turned down because they were a commercial and not a public operation. See Superintendant, Certificate Branch, to May–Gorman Airplanes, in M829 file 5, Glenbow Archives. See M829 as well for a share certificate for May–Gorman Airplanes, and an example of a commercial air pilot's certificate as issued by the Air Board. See B. D. Hobbs to May–Gorman Aeroplanes Ltd, 17 May 1920, and L. S. Breadner to May–Gorman Airplanes, 17 July 1920, as well as correspondence dated 10 June and 29 June 1920, for details on registering a plane under the new Air Board regulations.

See the City of Edmonton Archives, MS 56.6, for a copy of the agreement between the City of Edmonton and W. R. May and George W. Gorman to lease the "City of Edmonton" airplane.

For a general discussion of fairs and rodeos, see Donald G. Wetherell with Irene Kmet, *Useful Pleasures: The Shaping of Leisure in Alberta 1896–1945* (Regina, 1990), especially pages 311–41. See a letter from the Lloydminster Joint Agricultural Societies to W. J. Stark, Western Canada Fairs Association, 12 May 1919, in the City

Archie McMullen and Jock Palmer beside Great Western Airways' Stinson Detroiter, 1929. Courtesy Provincial Archives of Alberta, A12,031

of Edmonton Archives, Edmonton Exhibition Association Limited Collection, MS 3222, Class 9, S/c 4, file 2, and a general letter from May Airplanes Limited, 22 Jan 1920, in Glenbow Archives, M829, file 5, for details on exhibition flying.

On Ormer Locklear, see Art Ronnie, *Locklear: The Man Who Walked on Wings* (New York, 1973). For a slightly exaggerated American view of barnstorming and trick flying, see Don Dwiggins, *The Barnstormers: Flying Daredevils of the Roaring Twenties* (New York, 1968).

The best account of the 1920 cross-Canada flight sponsored by the Air Board can be found in F. H. Hitchins, *Air Board, Canadian Air Force, Royal Canadian Air Force*, Mercury Series, Canadian War Museum Paper No 2 (Ottawa, 1972), chapter IV. On Ernest Hoy and his flight over the Rockies, see Gowans, *Calgary* and *Lethbridge*; and Canadian Museum of Flight and Transportation and Lloyd M. Bungey, *Pioneering Aviation in the West as told by the Pioneers* (Surrey, BC, 1992), 41–3. See also, "Hoy Makes Historic Flight Over the Rockies," *Calgary Herald*, 13 Mar 1992.

For developments at High River Aerodrome, see Hitchins, *Air Board*; the Auditor General's *Reports* for the Air Board, 1919–20 to 1922, and the *Reports* for the Department of National Defence, Aviation sections, for 1923 through 1932. See also "High River: How the Government Protects Its Forest Reserves," *Imperial Oil Review* (April 1921), 18.

The story of the Rene and the Vic has been told many times. See, for example, Ellis, *Heritage*, 205–209; Elmer G. Fullerton and H. S. M. Kemp, "Making it Go with Guts and Glue, *True*, Canadian Edition (Mar 1960), 20A–20N; Fullerton published a personal account called "Pioneer Flying in the Canadian Sub-Arctic," *Imperial Oil Review* (Nov/Dec 1934), 26–30.

As always, these stories can be followed in the newspapers.

Out of the Shadows

For continuing developments at High River, see again Hitchins, *Air Board*; *Reports* for the Department of National Defence, Aviation sections; and the *High River Times*. The Auditor General's *Reports* have a wealth of detail on purchases, staff, and salaries. On forestry patrol work, see Peter J. Murphy, *History of Forest and Prairie Fire Control Policy in Alberta* (Edmonton, 1985), and Murphy, ed, *Recollections of a Young Forester in Alberta, 1920–1926* (Edmonton, 1990). The second work deals specifically with High River. See also Sidney P. Tucker, "Fire Ranging by Aeroplane in the Rockies," *Imperial Oil Review* (Feb 1923), 5–6, for details of patrol flights; and Tucker, "Forest Rangers of the Rockies," *Imperial Oil Review* (Mar 1925), 7–10. Readers should also consult Samuel Kostenuk and John Griffin, *RCAF Squadron Histories and Aircraft 1924–1968* (Toronto, 1977); and Larry Milberry, ed, *Sixty Years: The RCAF and CF Air Command 1924–1984* (Toronto, 1984).

Information on southern companies can be found in the newspapers, and in the indispensable Gowans's works *Calgary* and *Lethbridge*.

Developments in the Grande Prairie area can be followed in Myles, *Airborne;* •, the *Grande Prairie Herald*; and in J. G. MacGregor, *Grande Prairie* (Grande Prairie, 1983). For Edmonton, Myles, *Airborne*; and especially the *Edmonton*

Journal were invaluable. Record Group 11, Class 8 at the City of Edmonton Archives should also be consulted, as the progress of the airfield can be followed quite easily here. The letter from the Deputy Minister, Department of National Defence, to the City Clerk can be found here on file 1. See, as well, the summary in the "Edmonton Municipal Airport Souvenir Booklet Commemorating the Official Opening of the New Terminal, 24 Nov 1975 (MS 261-1, Class 1, file 1) in the City of Edmonton Archives. Edmund Dale, "The Role of Successive Town and City Councils in the Evolution of Edmonton, Alberta, 1892-1966" (PhD thesis, University of Alberta, 1969) concludes city councils played an important role in ensuring Edmonton developed as an aviation centre.

On Dalzell McKee, see the entries in Milberry, *Aviation;* and Sutherland, *Pioneers.* See, as well, AVM A. E. (Earl) Godfrey, "McKee Trans-Canada Flight," Canadian Aviation Historical Society *Journal* (Fall 1976), 90-94.

On the flying club movement in the late 1920s, see, for Edmonton, the newspapers; City of Edmonton Archives, RG 11, Class 8, and Class 7, file 2. See also Al Coyne, ". . . the Greatest Godam Flying School in the West" (University of Alberta Archives, 91-37, W. Max Fife clippings). For Calgary, see Gowans, *Calgary;* Patton Papers, Glenbow Archives M946; and the City Clerk's Papers at the City of Calgary Archives. See also the Souvenir Programme for the 28 Sep 1929 airport opening in Calgary in Fred Robert Gordon McCall Papers, Glenbow Archives M716; and G. L. Wilson, "Annual Air Show at Calgary Attracts Huge Crowd," *Canadian Aviation* (Nov 1929), 15, 36.

For what it was like to train with the clubs, see R. Murray Shortill, "Memories of Blatchford Field," Canadian Aviation Historical Society *Journal* (Winter 1979), 118-22; George Lothian, *Flight Deck: Memoirs of an Airline Pilot* (Toronto, 1979), 9; and John H. Blackburn, *Land of Promise* (Toronto, 1979), 181-82. The newspapers are also helpful, especially the "AeroTopics" column in the *Calgary Herald* during 1929-1930.

For the flying club story in Red Deer, the *Red Deer Advocate* was useful. I am indebted to Michael Dawe, Archivist, Red Deer and District Archives, for supplying additional information.

The journal *Canadian Aviation* regularly ran "News from the Flying Clubs" columns, and these are useful as well.

Information on aeronautical courses at the Provincial Institute of Art and Technology can be found in the institution's yearbooks and course calendars held by the library, Southern Alberta Institute of Technology, Calgary. Gowans, *Calgary*, also covers these events.

On airmail, see Milberry, *Aviation;* Georgette Vachon, *Goggles, Helmets, and Airmail Stamps* (Toronto, 1974); Manchester, *Industry*, 50-56; R. K. Malott, "Air Mail History," Canadian Aviation Historical Society *Journal* (Summer 1965), 43, 44, 46. Airmail activity is also carried in the Aviation sections of the Department of National Defence annual reports. The newspapers covered airmail developments carefully. The misfortunes of the Godfrey-Graham flight in the Peace River country can be followed in the *Peace River Times.*

On airport developments as a result of airmail in this period, see City of Edmonton Archives RG 11, Class 8; City Clerk's Papers at the City of Calgary Archives; Gowans, *Lethbridge*; and *Calgary*; annual reports from the Department of National Defence; and the newspapers.

Many details about Commercial Airways can be gleaned from the Wilfrid Reid "Wop" May Papers M829, at the Glenbow Archives, Calgary, including log books. See Commercial Airways Winter Schedule card for 1929–30 in the George William Scott Papers M1112, Box 1, file 6, Glenbow Archives, Calgary. See the Tariff list effective 1 May 1930 in the May Papers as well. See the letter from the Secretary Treasurer, James Richardson and Sons Limited, to Mayor Bury, 7 May 1928, for some details on Western Canada Airways planned expansion into Alberta, in City of Edmonton Archives RG 11, Class 8, file 1. The newspapers are also invaluable. For one account of Commercial's first official run down the Mackenzie, see Philip H. Godsell, *Pilots of the Purple Twilight* (Toronto, 1955), 113–50.

For southern developments, see Gowans, *Calgary* and *Lethbridge*; and Aviation History Reference notes at the Sir Alexander Galt Museum and Archives, Lethbridge. Retrospectives published in the newspapers to commemorate events are also useful. For this period, see the *Lethbridge Herald*, "Air Development Number," 3 June 1939.

There are many accounts of the May–Horner 1929 mercy flight. See, for example, Iris Allan, "Wop May, Leader of the First Mercy Flight," in City of Edmonton Archives, Bush Pilots Clipping File. A newspaper clipping from the *Suburban Times* (Edmonton) dated 2 Aug 1964 gives additional clarification on several details of the flight. The clipping can be found in Dr Harold A. Hamman Papers M2278, at the Glenbow Archives, Calgary. The *Edmonton Journal* followed every propeller turn. The *Peace River Record* accounts give a different perspective. Although grateful for the service rendered by the airplane and the flyers, the residents further north gave their greatest adulation to the mushers who brought the grim news from Red River on snowshoes at tremendous cost to their health. See *Peace River Record*, 25 Jan 1929, "Public Acknowledgement for Northern Mushers," and 1 Feb 1929, "Fortitude of La Fleur and Lambert Recognized with Presentation on Monday." See also the account in Gordon Reid, *Around the Lower Peace* (High Level, 1978), 19–21. For the letters mentioned in the text, see Helen MacMurchy to Captain W. R. May, 1 Feb 1929, and D. G. M. Queen to Mr May, 9 Jan 1929, in Wilfrid Reid "Wop" May Papers M829, file 7, Glenbow Archives, Calgary. Edmonton celebrations for May and Horner can be followed in the newspapers, and in RG 11, Class 8, file 26, in the City of Edmonton Archives.

For Fred McCall's nitroglycerine exploits, see the Calgary and Lethbridge papers, and "Fred McCall Air Adventurer" by Jack Peach, in Bush Pilots Clipping File, City of Edmonton Archives.

Sky Riders of the Plains

For the debate on airport expenditures, and the improvements themselves, see the *Edmonton Journal*, Jan to Dec 1929. See also the Edmonton Industrial Airport files, in RG 11, Class 8, at the City of Edmonton Archives; and Dale, "Role."

For the prairie airmail route, see the newspapers for the communities involved, and Gowans, *Calgary* and *Lethbridge*. See also Main, *Voyageurs*. Hitchins, *Air Board*, also gives it some attention. See, as well, the reports on civil aviation in Canada, in the Department of National Defence, Report of the Chief of General Staff, for the years ending 31 Mar 1931, and 31 Mar 1932.

Developments at High River can again be followed through Hitchins, *Air Board*; the *High River Times*; and the reports on civil aviation in the annual Department of National Defence reports.

The community newspapers are excellent sources of information on barnstorming in the early 1930s. See also R. Murray Shortill, "Barnstorming As I Saw It," Canadian Aviation Historical Society *Journal* (Summer 1970), 52–3. For a general discussion of leisure, see Wetherell with Kmet, *Useful Pleasures*; and as usual, see Gowans, *Calgary* and *Lethbridge*. For a discussion of the role of boards of trade in small town Alberta, see Donald G. Wetherell and Irene Kmet, *Town Life: Main Street and the Evolution of Small Town Alberta, 1880–1947*, forthcoming in the spring of 1995 from the University of Alberta Press in association with the Alberta Community Development and Alberta 2005 Centennial History Society. On boosterism, see Paul Voisey, *Vulcan: The Making of a Prairie Community* (Toronto, 1988), especially chapter 3, "Towns."

Accounts of the various tours in the newspapers can be supplemented with Milberry, *Aviation*; and Gowans, *Calgary* and *Lethbridge*. For the Ford Reliability Tours, see Ray Crone, "The Ford Reliability Air Tour of 1930," in Canadian Aviation Historical Society *Journal* (Spring 1978), 4–13; and Lesley Forden, *The Ford Air Tours, 1925–1931* (Nottingham, 1971). For the Trans-Canada Air Pageant, see Ray H. Crone, "The Trans-Canada Air Pageant of 1931," Canadian Aviation Historical Society *Journal* (Summer 1981), 35–48; and reports of the Royal Canadian Air Force in the Report of the Department of National Defence for the year ending 31 Mar 1932. Some details of both tours' stops in Edmonton can be found in RG 11, Class 8, files 4 and 5, City of Edmonton Archives. See, as well, the Edmonton Air Show *Official Directory* for the National Air Tour, 17 Sep 1930 in the Glenbow Library collection, Calgary.

Several general works on aviation history discuss air-mindedness, and the attraction of pioneering long distance flights. See, for example, Terry Gwynne-Jones, *Farther and Faster: Aviation's Adventuring Years 1909–1939* (Washington, 1991); and "Long-Distance Vision," in David McCullough, *Brave Companions: Portraits in History* (New York, 1992), 125–33. The *Edmonton Journal* is the best source for the long-distance flights that stopped in Edmonton, capturing the excitement as well as the details. For Post and Gatty, see, as well, "Around the World in Nine Days," *Imperial Oil Review* (April–June 1931), 1–2.

George Ross has been the subject of many newspaper and magazine articles.

See, for example, "Flying Rancher," *Maclean's*, 15 Sep 1948, by W. O. Mitchell, and "Flying Cowpokes of Lost River," in *Imperial Oil Review* (Sep 1959). The Ross Family Papers are held at the Glenbow Archives, Calgary.

Information on Dr Scott can be found in *Best in the West by a Damsite 1910–1940* (Bassano, 1974); on Joseph Austin, in *Memories of Ranfurly* (Ranfurly, 1983).

On Professor William Rowan, see the William Rowan Papers, University of Alberta Archives, Edmonton, and "The Memorable William Rowan," *New Trail* (Summer 1993), 13–17.

On the Alberta Provincial Police, see Sean Moir, "The Alberta Provincial Police, 1917–1932" (MA thesis, University of Alberta, 1992).

Commercial developments in southern Alberta in the 1930s can be pieced together from Gowans and the newspapers. Collection P19911057126 at the Sir Alexander Galt Museum and Archives in Lethbridge contains many details on flying in the south. Florence Whyard's *Ernie Boffa: Canadian Bush Pilot* (Alaska, 1984) is very good on flying conditions in southern Alberta during this period. Z. L. Leigh's hardnosed memoir *And I Shall Fly* (Toronto, 1985) shows just how difficult it was to make a living from flying. See Les Stahl, *A Record of Service: The History of Western Canada's Pioneer Gas and Electric Utilities* (Edmonton, 1987) for an example of a company's use of planes.

From Sea to Sea to Sea
The W. R. "Wop" May Papers M829, file 8, at the Glenbow Archives, has a copy of the commemorative menu card from the Testimonial Dinner for the Grads, Dickins, and May.

For general information on the Canadian Shield and prospecting down north, see Morris Zaslow, *The Northward Expansion of Canada, 1914–1967* (Toronto, 1988), chapter 4; and Zaslow, *Reading the Rocks: The Story of the Geological Survey of Canada 1842–1972* (Ottawa, 1975). On the Northern Aerial Minerals Exploration Company, see Sutherland, *Pioneers*, 93. On the development of aerial photography, see Don W. Thompson, *Skyview Canada: A Story of Aerial Photography in Canada* (Ottawa, 1975).

For the story of Doc Oaks and the Elliott brothers, see Sutherland, *Pioneers*, 23.

There is a great deal of information available on companies and individuals flying down north. K. M. Molson's *Pioneering in Canadian Air Transport* (Winnipeg, 1974) is particularly good because of the detail it contains. William Paul Ferguson's *The Snowbird Decades* (Vancouver, 1979) is also helpful. Many details can be gleaned from various entries in Sutherland's *Pioneers*.

Several biographies and memoirs are full of details on the challenges and experiences of flying in northern Alberta and beyond. Leigh's memoirs and Whyard's *Boffa*, both listed previously, are good sources. So is Ronald A. Keith's *Bush Pilot with a Briefcase: The Happy-go-Lucky Story of Grant McConachie* (Toronto, 1972). See also Walter Henry and the Canadian Bush Pilot Project, *Uncharted Skies: Canadian Bush Pilot Stories* (Calgary, 1983); J. A. Foster, *The Bush Pilots: A Pictorial History of a Canadian Phenomenon* (Toronto, 1990); Walter E. Gilbert and Kathleen

Shackleton, *Arctic Pilot: Life and Work on North Canadian Air Routes* (Toronto, 1940); Iris Allen, *Wop May, Bush Pilot* (Toronto, 1966); and "Bush Pilots: A Pictorial Feature," *Alberta History* (Summer 1983), 14–21. See "Flights and Fights of a Pioneer Pilot," and "Lucky Lucas" in *Flying the Frontiers: A Half-million Hours of Aviation Adventure* (Saskatoon, 1994) by Shirlee Smith Matheson for good portrayals of many aspects of early aviation in Alberta, including northern flying. See also Duncan D. McLaren, *Bush to Boardroom: A Personal View of Five Decades of Aviation History* (Winnipeg, 1992).

Log books are valuable sources, as they record the particulars of every flight taken in a plane. See, for example, the log book for Fokker Universal G-CAHE in the Stanley N. Green Papers M454 at the Glenbow Archives; and the log book for Bellanca Pacemaker CF-AR1 in the W. R. "Wop" May Papers M829, file 4, in the Glenbow Archives. See, as well, typed sheets listing what was done during the overhaul of CF-ATJ, CF-ATW, and CF-AUD in Nov 1933, found in the A. M. Berry Papers 72.295, file 11, at the Provincial Archives of Alberta. The R. M. Shortill collection of photographs at the PAA, accession 73.549, is also an excellent resource.

Many other sources also provide useful information. Airline schedules provide route and rate details. See Mackenzie Air Service Schedules and Tariffs in the Glenbow Library. Because of its stake in the development of aviation as a viable transportation technology, Imperial Oil published triumphs and developments in the *Review*. See, for example, T. H. Inkster, "Fairbairn of Waterways," *Imperial Oil Review* (April/May 1936), 29; and W. B. Burchall, "The Flying Boxcar," *Imperial Oil Review* (Jan/Feb 1933), 18–20. The *Nor'West Miner* was a tireless booster of developing transportation and industry down north. From ads to articles, it's invaluable.

On mining developments down north in the 1930s, see the *Nor'West Miner*, all issues. The *Bulletins* put out by the Canadian Institute of Mining are also useful, as are articles in the *Canadian Mining Journal*. See, for example, in the *Journal*, Hugh G. Spence, "Status of Mining Developments for Silver and Radium in Northwestern Canada, 1935," in the Dec 1935 issue. On the discoveries and subsequent development at Great Bear Lake, see Gilbert and Shackleton, *Arctic Pilot*, chapter 8. The definitive study is certainly Robert Bothwell's *Eldorado: Canada's National Uranium Company* (Toronto, 1984). See, as well, F. Turley, "The Trail of 32," *Imperial Oil Review* (May/June 1932), 16–17. Frederick B. Watt's *Great Bear: A Journey Remembered* (Yellowknife, 1980) is an exciting account by someone who was there. The newspapers reported developments frequently. For just two examples from the *Edmonton Journal*, see "Northland Flying on Record Scale by Airways Companies," 3 May 1930, and "Air Transportation Indispensable to Northern Development," 18 May 1937.

On carrying canoes by plane, see J. J. Green, "Effect on Performance of Carrying a Canoe on an Aeroplane" (National Research Council Report PAA-24, Ottawa, 1935); and Sutherland, *Pioneers*, 95.

On fish hauling, see Keith, *Bush Pilot*; and "Flights and Fights of a Pioneer Pilot," in Matheson, *Flying*. A great deal was pieced together from the *Grande Prairie Herald* and *Peace River Record*. Local histories are also useful. See, for example,

Treasured Scales of the Kinosoo (Cold Lake, 1980) for snippets that can be located in family histories on winter fishing and fish hauling in the Cold Lake area.

Aviation developments in Edmonton and Cooking Lake during the 1930s got good press coverage in the *Edmonton Journal*. The *Nor'West Miner* was also an Edmonton booster and carried lots of information about aviation in Edmonton. Airport developments and relations with the flying club can be followed at the City of Edmonton Archives, RG 11, Class 8. For Cooking Lake, see also Les Faulkner, *Wheels, Skis and Floats: A History of the Cooking Lake Seaplane Base* (Edmonton, 1992). Edmund Dale, "Role," also contains good information.

Information on developments in Calgary comes from the *Calgary Herald*; Gowans, *Calgary*; and in the annual reports of the Municipal Airport held by the City of Calgary Archives.

On developments in the Peace River–Grande Prairie area, see the *Northern Gazette* and the *Grande Prairie Herald*. See also Isabel M. Campbell, *Grande Prairie: Capital of the Peace* (Grande Prairie, 1968); and MacGregor, *Grande Prairie*. "Lucky Lucas," in Matheson, *Flying*, also contains some interesting reminiscences about flying for Peace River Airways in the late 1930s.

On Trans-Canada Air Lines, a good place to start is Robert Bothwell and William Kilbourne, *C. D. Howe: A Biography* (Toronto, 1979), chapter 8, "My Airline." George Lothian, *Flight Deck: Memoirs of an Airline Pilot* (Toronto, 1979) should also be consulted, as should David H. Collins, *Wings Across Time: The Story of Air Canada*

Captain Fred McCall, World War I Ace, photographed with a Curtiss Jenny JN–4D at Victoria Park, 1919. Courtesy Glenbow Archives, Calgary, NA–1258–22

(Griffin House, 1978). Molson, *Pioneering,* is important for James Richardson's Canadian Airways perspective. See also "Philip Johnson, 1938," in Sutherland, *Pioneers,* 108–14; and MacLaren, *Bush,* chapter 4. See also "The Trans-Canada Airway and its Relationship to the World's Airway System," in *Canada Yearbook 1938,* 11–19 plus map.

The newspapers are the best sources for local developments and reactions. Gowans should be consulted for developments in Calgary and Lethbridge. For Lethbridge, as well, see the *Herald's* "Special Aviation Number" published to mark the opening of Kenyon airfield; and the *Souvenir Programme* from the opening of Kenyon Field in the J. D. Higinbotham Papers M517, file 33, in the Glenbow Archives. Reference material in P19911057127 at the Sir Alexander Galt Museum and Archives in Lethbridge contains information on the radio range station.

RG 11, Class 8, at the City of Edmonton Archives is an invaluable source once again for everything from the initial lobbying of the federal government for help with airport expenditures to the agreement with TCA and expanded airport facilities. See, as well, RG 11, Class 8 and 22, and RG 8, Class 32. RG 11 Class 15, file 44 contains TCA schedules for 1939 and 1940.

The World In Our Back Yard

Two books that should be consulted for general background to World War II in Canada are Bothwell et al, *Canada 1900–1945*; and J. L. Granatstein and Desmond Morton, *A Nation Forged in Fire: Canadians and the Second World War 1939–1945* (Toronto, 1989). For Alberta, see Palmer, *Alberta.* For background on small town life in Alberta, see Wetherell and Kmet, *Town Life.*

For the RCAF see W. A. B. Douglas, *The Creation of a National Air Force: The Official History of the Royal Canadian Air Force,* Vol II (Toronto, 1986); Brereton Greenhous et al, *The Crucible of War: The Official History of the Royal Canadian Air Force,* Vol III (Toronto, 1994); Kostenuk and Griffin, *RCAF*; and Milberry, *Sixty Years,* 97–108.

Many works discuss the British Commonwealth Air Training Plan from its genesis to its closure. See *Aerodrome of Democracy: Canada and the British Commonwealth Air Training Plan, 1939–1945* (Ottawa, 1983) by F. J. Hatch for a discussion of all aspects of the plan. Readers should also consult Ted Barris, *Behind the Glory: The Plan that Won the Allied Air War* (Toronto, 1992); Larry Milberry and Hugh A. Halliday, *The Royal Canadian Air Force at War, 1939–1945* (Toronto, 1990); Spencer Dunmore, *Wings for Victory: The Remarkable Story of the British Commonwealth Air Training Plan in Canada* (Toronto, 1994); Peter Conrad, *Training For Victory* (Saskatchewan, 1989); Mary Ziegler, *We Serve that Men May Fly: The Story of Women's Division, Royal Canadian Air Force* (Hamilton, 1973); and Brereton Greenhous and Norman Hillmer, "The Impact of the British Commonwealth Air Training Plan on Western Canada: Some Saskatchewan Examples," *Journal of Canadian Studies* (Fall/Winter 1981), 133–44.

The negotiations to formulate the plan can be followed in J. L. Granatstein, *Canada's War: The Politics of the Mackenzie King Government, 1939–1945* (Toronto, 1975), 43–66; and J. W. Pickersgill, *The Mackenzie King Record*, Vol 1, 1939–1944 (Toronto, 1960), 40–59.

For the experiences of those in the plan in Alberta, see Murray Peden, *A Thousand Shall Fall* (Toronto, 1988). Peden trained at Edmonton and High River. John W. Chalmers has recorded his experiences as an air observer trainee in "Learning the Gen Trade," *Alberta History* (Summer 1994), 2–10. Gordon Wagner details his stay in Calgary and Fort Macleod in *How Papa Won the War* (Courtney, BC, 1989). Log books are interesting records of training. See, for example, *Observer's and Air Gunner's Flying Log Book RCAF*, in the William E. Rowbotham Papers M1076 at the Glenbow Archives.

The contemporary newspapers are full of information on the stations and activities. Many of the papers in the small towns put out special editions to commemorate a school's opening. See, for example, *Vulcan Advocate*, 5 Nov 1942. More recently, retrospectives and reminiscences have been published See, for example, "War-time romance with 'cute' RAF man blossomed into a lifetime of happiness," *Edmonton Journal*, 11 Oct 1994; and "Flight School Remembered," The *Medicine Hat News*, 10 Nov 1989. Daily diaries were kept at each station, and they are now housed at the National Archives in Ottawa. Most stations had newsletters, and put out highly descriptive programs on anniversary dates. See, for example, *Souvenir Programme, No 5 EFTS*, Third Anniversary, 4 Sep 1944, in the Glenbow Archives. See, as well, copies of the *Penhold Log* in the Red Deer and District Archives. *A History of No 5 EFTS* (Feb 1945) at the Glenbow Library was also helpful. See also "Bowden Field Number 32 Elementary Flying Training School," in *Pioneer Legacy: Bowden and Districts* (Bowden, 1979), 89–91; and "The Vulcan Airport," in *Wheat Country: A History of Vulcan and District I* (Vulcan and District Historical Society, 1973), 105–106.

For the role of the Legion in establishing recreational facilities, see Clifford H. Bowering, *Service: The Story of the Canadian Legion* (Ottawa, 1960). The article on John Armistead "J. A." Wilson in Sutherland, *Pioneers*, gives a good, short introduction to the BCATP.

For information on the Edmonton and Northern Alberta Aero Club initial arrangements to train air crew in 1939, see RG 11, Class 8, file 10 in the City of Edmonton Archives. For the response to Croil's recommendations, see RG 11, Class 8, file 16 in the City of Edmonton Archives. For the delivery of American planes to the RAF in Nov and Dec 1939, see *Lethbridge Herald*, 18, 20, 27 Nov and 15, 18, 21, 28 Dec; and *The History of the Border County of Coutts, 1890–1965* (Coutts, 1965), 299–300.

For an example of a city adopting an air squadron, see the City Clerk's Papers, Box 363, file 2375; and Box 352, file 2294, City of Calgary Archives. For developments at the Calgary airport, see the *Annual Reports* for the Municipal Airport, in Annual Reports, Boxes 283 and 284; and the *Annual Reports* of the City of Calgary Electric Light and Power Department, in RG 1, Box 282, file 1954, City of Calgary Archives.

On the defence projects in northwestern Canada, see Zaslow, *Northward Expansion*, chapter 8; Colonel Stanley W. Dziuban, *Military Relations Between the United States and Canada 1939–1945* (Washington, DC, 1959); Stacey, *Arms*, 379–92; Stan Cohen, *The Forgotten War: A Pictorial History of World War II in Alaska and*

Northwest Canada (Missoula, MT, 1981); D. B. Wallace, "Canada's Northern Air Routes," *Canadian Geographical Journal* (Oct 1943), 186–201; Wesley Frank Craven and James Lea Cate, eds, *The Army Air Forces in World War II Services Around the World*, Vol 7 (Chicago, 1950); Bob Hesketh, ed, *Three Northern Wartime Projects: The Northwest Staging Route, Alaska Highway and CANOL* (forthcoming, Spring 1995, University of Alberta Press); and Hope Morritt, *Land of the Fireweed: A Young Woman's Story of Alaska Highway Construction Days* (Edmonds, WA, 1985).

Two significant books on the Alaska Highway that should be consulted by any researcher are Kenneth Coates and W. R. Morrison, *The Alaska Highway in World War II: The U. S. Army of Occupation in Canada's Northwest* (Toronto, 1992), and Heath Twichell, *Northwest Epic: The Building of the Alaska Highway* (New York, 1992).

On the Northwest Staging Route, see Charles F. O'Brien, "Northwest Staging Route," *Alberta Historical Review* (Autumn 1969), 14–22; J. A. Wilson, "Northwest Passage by Air," *Canadian Geographical Journal* (Mar 1943), 107–130. See also Major Edwin R. Carr, *History of the Northwest Air Route to Alaska 1942–1945*, in the Major Edwin R. Carr Papers M3602, Glenbow Archives.

On the CANOL project, see the interpretively flawed but detailed *The CANOL Project: An Adventure of the U.S. War Department in Canada's Northwest* (Edmonton, 1985), by P. S. Barry; and Oliver B. Hopkins, "The 'CANOL' Project," *Canadian Geographical Journal* (Nov 1943), 238–49.

Some details of changes at Edmonton's airport can be found in RG 11, Class 8, File 16.

Archie McMullen's log book detailing test flights at Aircraft Repair Ltd is in Accession 84.27, the McMullen Papers at the Provincial Archives of Alberta. The Torrance Papers, Acc 77.3, also at the PAA, contain many details of operations at Aircraft Repair Ltd during the war. See also John Gilpin, *Edmonton: Gateway to the North* (Edmonton, 1984), 170–72.

On the development of radio communications, the leading role played by Mackenzie Air Service can be followed in Ferguson, *Snowbird*, 47–76. The most detailed account for the war years is Norman Larson, *Radio Waves Across Canada and Up the Alaska Highway*, Occasional Paper No 25 (Lethbridge Historical Society, 1992). On meteorology, see Dr Donald B. Kennedy, "Fifty Years of Aviation Meteorology," Canadian Aviation Historical Society *Journal* (Fall 1980), 78–82.

On Canadian Pacific, see Keith, *Bush Pilot*, chapter 26; Sutherland, *Pioneers*, 155–60; "Historical Highlights, and the Men Behind the Line," *Canadian Aviation* (June 1977), 43–45; John Blatherwick, *A History of Airlines in Canada* (Toronto, 1989); and D. M. Bain, *Canadian Pacific Air Lines: Its History and Aircraft* (Calgary, 1987).

Epilogue: One Day Wide
See the TCA and CP ads in various issues of the *Canadian Geographical Journal*, for example. The Southern Alberta Aviation Conference can be followed in the *Calgary Herald*.

On Canada's role in immediate postwar international aviation, see David Mackenzie, *Canadian and International Civil Aviation 1932–1948* (Toronto, 1989), chapters 7 through 11; and Main, *Voyageurs*, 187–219.

For excellent records on postwar aviation committees, see the Chamber of Commerce Collection, M86.9, files 51 and 52, Medicine Hat Museum and Art Gallery Archives; and the City of Calgary Archives, Special Aviation Commission Papers, RG1515. Annual reports are also useful, as RG 11, Class 8, files 13, 19, 24, and 25 at the City of Edmonton Archives.

For examples of companies that started up right after the war, see Sutherland, *Pioneers*, 100, 276–77. See the Tommy Fox files at Canada's Aviation Hall of Fame, Wetaskiwin, for the Associated Airways story. On Westland Spraying Service, see Matheson, *Flying*, 118–21.

On company planes, see, for example, "Business on the Wing," *Imperial Oil Review* (Aug/Sep 1949), 21–25.

On postwar flying clubs, see, for example, Sutherland, *Pioneers*, 188–92; "Schools in the Sky," *Imperial Oil Review* (Jan 1949), 34–37. Club records and local histories are very good sources, too. See, for example, *In the Beginning: A History of Coronation, Throne, Federal and Fleet Districts* (Coronation, 1979), 26.

On search and rescue, see "Wop" May Scrapbook, Accession 87.6, at the Provincial Archives of Alberta; Milberry, *Sixty Years*, 209; and *Circulars to Civilian Pilots*, "The Royal Canadian Air Force Search and Rescue Service," dated 2/11/48. On immediate postwar adjustment for the RCAF, see Milberry, *Sixty Years*, 196–98; and Arnold P. Vaughan, *418 City of Edmonton Squadron History* (Edmonton, 1984).

On the RCMP air division, see Arnold P. Vaughan, *RCMP "Air" Division 1937–1973* (Ottawa, 1973); and E. P. Gardiner, "Winged Mounties," Canadian Aviation Historical Society *Journal* (Winter 1968), 104–107.

On Leigh Brintnell and Northwest Industries, see Reynolds Aviation Museum research files, including an interview with Stan McMillan, 17 Jan 1991; Peter Marshall, "Historic Bellanca is still in the Running," *Canadian Aviation* (May 1969), 26–27, 48; "Sky rocket Bush Plane Makes Debut in Western Canada," *Canadian Aviation* (Mar 1946), 35, 70. See the ad for Northwest's Bellanca in *Canadian Aviation* (Jan 1947), 60.

Index

(Numbers in bold indicate photographs)